my **revisi⏻n** notes

AQA A-level
PHYSICS

Keith Gibbs

HODDER
EDUCATION
AN HACHETTE UK COMPANY

Hachette UK's policy is to use papers that are natural, renewable and recyclable products and made from wood grown in sustainable forests. The logging and manufacturing processes are expected to conform to the environmental regulations of the country of origin.

Orders: please contact Bookpoint Ltd, 130 Park Drive, Milton Park, Abingdon, Oxon OX14 4SE. Telephone: (44) 01235 827720. Fax: (44) 01235 400454. Email education@bookpoint.co.uk Lines are open from 9 a.m. to 5 p.m., Monday to Saturday, with a 24-hour message answering service. You can also order through our website: www.hoddereducation.co.uk

ISBN: 978 1 4718 5479 8

© Keith Gibbs 2017

First published in 2017 by
Hodder Education,
An Hachette UK Company
Carmelite House
50 Victoria Embankment
London EC4Y 0DZ
www.hoddereducation.co.uk

Impression number   10 9 8 7 6 5 4 3 2 1

Year   2021 2020 2019 2018 2017

Cover photo
Typeset in Bembo Std Regular 11/13 by Integra Software Services Pvt. Ltd., Pondicherry, India
Printed in Spain

A catalogue record for this title is available from the British Library.

Get the most from this book

Everyone has to decide his or her own revision strategy, but it is essential to review your work, learn it and test your understanding. These Revision Notes will help you to do that in a planned way, topic by topic. Use this book as the cornerstone of your revision and don't hesitate to write in it — personalise your notes and check your progress by ticking off each section as you revise.

Tick to track your progress

Use the revision planner on pages 4 and 5 to plan your revision, topic by topic. Tick each box when you have:

- revised and understood a topic
- tested yourself
- practised the exam questions and gone online to check your answers and complete the quick quizzes

You can also keep track of your revision by ticking off each topic heading in the book. You may find it helpful to add your own notes as you work through each topic.

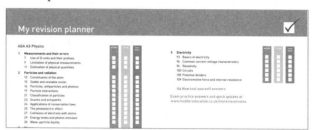

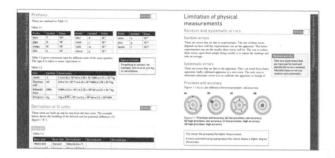

Features to help you succeed

Exam tips

Expert tips are given throughout the book to help you polish your exam technique in order to maximise your chances in the exam.

Typical mistakes

The author identifies the typical mistakes candidates make and explains how you can avoid them.

Now test yourself

These short, knowledge-based questions provide the first step in testing your learning. Answers are at the back of the book.

Definitions and key words

Clear, concise definitions of essential key terms are provided where they first appear.

Key words from the specification are highlighted in bold throughout the book.

Revision activities

These activities will help you to understand each topic in an interactive way.

Exam practice

Practice exam questions are provided for each topic. Use them to consolidate your revision and practise your exam skills.

Summaries

The summaries provide a quick-check bullet list for each topic.

Online

Go online to check your answers to the exam questions and try out the extra quick quizzes at **www.hoddereducation.co.uk/myrevisionnotes**

My revision planner

REVISED   TESTED   EXAM READY

Exam practice answers and quick quizzes at **www.hoddereducation.co.uk/myrevisionnotes**

REVISED TESTED EXAM READY

Exam practice answers and quick quizzes at
www.hoddereducation.co.uk/myrevisionnotes

Countdown to my exams

6–8 weeks to go

- Start by looking at the specification — make sure you know exactly what material you need to revise and the style of the examination. Use the revision planner on pages 4 and 5 to familiarise yourself with the topics.
- Organise your notes, making sure you have covered everything on the specification. The revision planner will help you to group your notes into topics.
- Work out a realistic revision plan that will allow you time for relaxation. Set aside days and times for all the subjects that you need to study, and stick to your timetable.
- Set yourself sensible targets. Break your revision down into focused sessions of around 40 minutes, divided by breaks. These Revision Notes organise the basic facts into short, memorable sections to make revising easier.

REVISED

2–6 weeks to go

- Read through the relevant sections of this book and refer to the exam tips, exam summaries, typical mistakes and key terms. Tick off the topics as you feel confident about them. Highlight those topics you find difficult and look at them again in detail.
- Test your understanding of each topic by working through the 'Now test yourself' questions in the book. Look up the answers at the back of the book.
- Make a note of any problem areas as you revise, and ask your teacher to go over these in class.
- Look at past papers. They are one of the best ways to revise and practise your exam skills. Write or prepare planned answers to the exam practice questions provided in this book. Check your answers online and try out the extra quick quizzes at **www.hoddereducation.co.uk/myrevisionnotes**
- Use the revision activities to try out different revision methods. For example, you can make notes using mind maps, spider diagrams or flash cards.
- Track your progress using the revision planner and give yourself a reward when you have achieved your target.

REVISED

One week to go

- Try to fit in at least one more timed practice of an entire past paper and seek feedback from your teacher, comparing your work closely with the mark scheme.
- Check the revision planner to make sure you haven't missed out any topics. Brush up on any areas of difficulty by talking them over with a friend or getting help from your teacher.
- Attend any revision classes put on by your teacher. Remember, he or she is an expert at preparing people for examinations.

REVISED

The day before the examination

- Flick through these Revision Notes for useful reminders, for example the exam tips, topic summaries, typical mistakes and key terms.
- Check the time and place of your examination.
- Make sure you have everything you need — extra pens and pencils, tissues, a watch, bottled water, sweets.
- Allow some time to relax and have an early night to ensure you are fresh and alert for the examinations.

REVISED

My exams

Physics Paper 1

Date:..

Time:...

Location:...

Physics Paper 2

Date:..

Time:...

Location:...

Physics Paper 3

Date:..

Time:...

Location:...

1 Measurements and their errors

Use of SI units and their prefixes

Fundamental (base) units

Mass — measured in kilograms

The kilogram (kg) is the mass equal to that of the international prototype kilogram kept at Sevres, France.

Length — measured in metres

The metre (m) is the distance travelled by electromagnetic waves in free space in $1/299\,792\,458\,s$.

Time — measured in seconds

The second (s) is the duration of $9\,192\,631\,770$ periods of the radiation corresponding to the transition between two hyperfine levels of the ground state of caesium-137 atom.

Further SI units

Electric current — measured in amperes

The ampere (A) is that constant current that, if maintained in two parallel straight conductors of infinite length and of negligible circular cross section placed 1 metre apart in a vacuum, would produce a force between them of $2 \times 10^{-7}\,N$.

Temperature — measured in kelvin

The kelvin (K) is $1/273.16$ of the thermodynamic temperature of the triple point of water.

Amount of substance — measured in moles

The mole (mol) is the amount of substance in a system that contains as many elementary particles as there are in $0.012\,kg$ of carbon-12.

> **Exam tip**
>
> Remember to use the appropriate SI units.

Prefixes

These are outlined in Table 1.1.

Table 1.1

Prefix	Symbol	Value
tera	T	10^{12}
giga	G	10^{9}
mega	M	10^{6}
kilo	k	10^{3}

Prefix	Symbol	Value
deci	d	10^{-1}
centi	c	10^{-2}
milli	m	10^{-3}
micro	μ	10^{-6}

Prefix	Symbol	Value
nano	n	10^{-9}
pico	p	10^{-12}
femto	f	10^{-15}

Table 1.2 gives conversion rates for different units of the same quantity. The sign ≡ is taken to mean 'equivalent to'.

Typical mistake

Forgetting to convert, for example, mm to m or g to kg in calculations.

Table 1.2

Unit	Symbol	Conversions
Joule	J	$1\,J \equiv 6.24 \times 10^{18}\,eV \equiv 2.78 \times 10^{-7}\,kWh \equiv 1.11 \times 10^{-17}\,kg$
Electron volt	eV	$1\,eV \equiv 1.6 \times 10^{-19}\,J \equiv 4.45 \times 10^{-26}\,kWh \equiv 1.78 \times 10^{-36}\,kg$
Kilowatt hour	kWh	$1\,kWh \equiv 3.6 \times 10^{6}\,J \equiv 2.25 \times 10^{25}\,eV \equiv 4.01 \times 10^{-11}\,kg$
Kilogram	kg	$1\,kg \equiv 8.99 \times 10^{16}\,J \equiv 5.6 \times 10^{35}\,eV \equiv 2.5 \times 10^{10}\,kWh$

Derivation of SI units

These units are built up step by step from the base units. The example below shows the building of the derived unit for potential difference (V) ($kg\,m^2\,s^{-3}\,A^{-1}$).

Example

Table 1.3

Base unit	Base unit	Derived unit	Derived unit	Derived unit
Metre (m)	Second	Velocity ($m\,s^{-1}$)		
Ampere (A)	Second	Charge (C) ($A\,s$)		
Second (s)		Velocity	Acceleration ($m\,s^{-2}$)	
Kilogram (kg)		Acceleration	Force (N) ($kg\,m\,s^{-2}$)	
	Metre	Force	Work (J) ($kg\,m^2\,s^{-2}$)	
		Work	Charge	Potential difference (V) ($kg\,m^2\,s^{-3}\,A^{-1}$)

Now test yourself

1 What is the result of multiplying 10 MN by 25 pm? Your answer should include both the numerical answer and the correct unit.
2 Using a table like the one above, show the building of the derived unit Pascal (Pa).

Answers on p. 216

Limitation of physical measurements

Random and systematic errors

Random errors

These are errors that are due to experimenter. The size of these errors depends on how well the experimenter can *use* the apparatus. The better experimenter you are the smaller these errors will be. The way to reduce these errors, apart from simply being careful, is to repeat the readings and take an average.

Systematic errors

These are errors that are due to the apparatus. They can result from faulty apparatus, badly calibrated apparatus or a zero error. The only way to eliminate systematic errors is to re-calibrate the apparatus or change it!

> **Revision activity**
>
> Take one experiment that you have performed and identify the errors involved. Tabulate these errors as random and systematic.

Precision and accuracy

Figure 1.1 shows the difference between **precision** and **accuracy**.

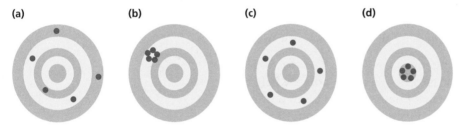

(a) **(b)** **(c)** **(d)**

Figure 1.1 Precision and accuracy: (a) low precision, low accuracy; (b) high precision, low accuracy; (c) low precision, high accuracy; (d) high precision, high accuracy

> The closer the grouping the higher the **precision**.
>
> A more symmetrical grouping about the centre shows a higher degree of **accuracy**.

Measurements can also be considered in terms of their **repeatability** (whether they can be repeated), their **reproducibility** (whether their values can be reproduced when measured many times) and their **resolution** (an example of resolution would be pixels per mm² in an image).

Uncertainty

The uncertainty (ΔQ) in a quantity Q ($Q = a + b$) is:

$$\Delta Q = \Delta a + \Delta b$$

where Δa and Δb are the uncertainties in the quantities a and b. The percentage uncertainty ($\%Q = (\Delta Q/Q) \times 100$) is:

$$\%Q = \%a + \%b$$

If $Q = anb$ (where n can be any number including 1):

$$\Delta Q = b\Delta a + an\Delta b$$

and

$$\%Q = \%a + n\%b$$

Find the maximum possible percentage uncertainty in the measurement of the acceleration of an object that moves at $20 \pm 1 \text{ m s}^{-1}$ in a circle of radius 5 ± 0.2 m. ($a = v^2/r$)

Answer

$$\%a = (2 \times \%v) + \%r = \left[\left(2 \times \frac{1}{20}\right) + \left(\frac{0.2}{5}\right)\right] \times 100 = 14\%$$

But:

$$a = \frac{20^2}{5} = 80 \text{ m s}^{-2}$$

Therefore the answer for a should be quoted as:

acceleration $(a) = 80 \text{ m s}^{-2} \pm 14\%$

Uncertainty in graphs

The uncertainty in any point on a graph is shown by the error bars.

Figure 1.2 shows a series or readings of voltage and current for a metal wire. The line of gradient m is the best-fit line to the points where the two extremes, m_1 and m_2 show the maximum and minimum possible gradients that still lie through the error bars of all the points. The percentage uncertainty in the gradient is given by:

$$\frac{m_1 - m_2}{m} = \left(\frac{\Delta m}{m}\right) \times 100\%$$

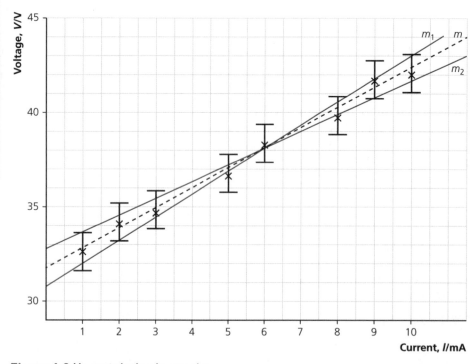

Figure 1.2 Uncertainties in graphs

Example

In the example above:

$$m_1 = \frac{(43.2 - 30.8)}{0.01} = 1240\,\Omega$$

and

$$m_2 = \frac{(41.7 - 32.7)}{0.01} = 900\,\Omega$$

gradient of the best-fit line $(m) = \frac{(42.4 - 31.8)}{0.01} = 1060$

In the example the uncertainty in the gradient (resistance) is:

$$\frac{(1240 - 900)}{1060} = \pm 32\%$$

Alternatively the value of the gradient (resistance) can be written as $1060 \pm 340\,\Omega$.

Estimation of physical quantities

Orders of magnitude

REVISED

Physicists use the phrase 'the right order of magnitude' to refer to a number in the right sort of range. For example, finding the time of swing of a 1-metre pendulum as 1.2 s and not 12 s, the specific heat capacity of water as $4500\,\mathrm{J\,kg^{-1}\,K^{-1}}$ and not $45\,000\,\mathrm{J\,kg^{-1}\,K^{-1}}$, or working out that the refractive index of an air–glass interface is 1.4 and not 0.4.

Estimation of approximate values of physical quantities

REVISED

It is always a good idea to be able to estimate the size of a quantity, so that when you work out a problem or finish an experiment you have a rough idea of what sort of value to expect.

Exam practice

1 Using a table similar to Table 1.3 (p. 8) show that the derived unit for resistance (the ohm) can be expressed in SI base units as $\mathrm{kg\,m^2\,A^{-2}\,s^{-3}}$. [3]
2 The derived SI unit for work is:
   A watt
   B joule per second
   C newton second
   D joule. [1]
3 The resistance of a 60 cm length of wire is $0.5\,\Omega$ and its diameter 0.3 mm. If the uncertainty in the measurement of its length is 5%, that of the diameter 2% and that of the resistance 8%, calculate the resistivity of the wire and give the percentage accuracy of your answer. [3]
4 The density of a spherical ball of iron is measured by finding its mass and then measuring its diameter. The mass can be measured to ± 10 g and the diameter to ± 2 mm. If the mass of the ball is found to be 1.54 kg and the diameter 7.2 cm, which of the following is closest to the correct accuracy for its density?
   A $\pm 270\,\mathrm{kg\,m^{-3}}$
   B $\pm 700\,\mathrm{kg\,m^{-3}}$
   C $\pm 270\,\mathrm{g\,m^{-3}}$
   D $\pm 700\,\mathrm{g\,m^{-3}}$ [1]

5 Figure 1.3 shows a series of readings of applied force and length for a metal wire.

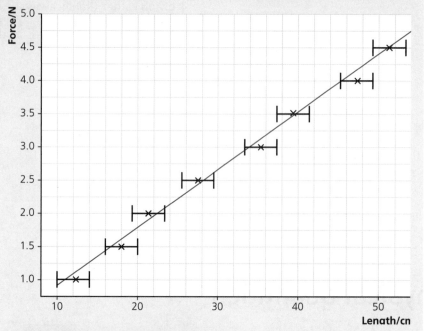

Figure 1.3 **Extension of a metal wire**

Use the graph to find:
(a) the value of length when the applied force is zero [3]
(b) the percentage uncertainty in the value of the length of the wire when the applied force is 30 N [2]
(c) the mean value for the gradient of the line [2]
(d) the uncertainty in the value that you calculated. [2]

Answers and quick quiz 1 online

ONLINE

Summary

You should now have an understanding of:
● fundamental (base) units — kilogram, metre, second
● how to derive further SI units, such as potential difference, resistance, momentum and pressure
● prefixes — tera, giga, mega, kilo, centi, milli, micro, nano, pico and femto
● how to convert between units
● random and systematic errors — random errors are due to the experimenter and systematic errors are due to the apparatus
● repeatability, reproducibility, resolution and accuracy — precision is shown by a close grouping of results and accuracy by a symmetrical grouping
● uncertainty in measurements — the uncertainty of a compound quantity can be found by adding the uncertainties of its parts, whether fractional or percentage
● uncertainty in graphs — add error bars to points on a graph and then draw the best-fit line through the spread of points
● orders of magnitude — the size of a quantity within a factor of ten
● the importance of estimation — the ability to predict the approximate order of magnitude of a quantity

2 Particles and radiation

Constituents of the atom

Protons, neutrons and electrons

It was the experiments on the scattering of alpha particles by gold nuclei in the early part of the twentieth century that laid the foundation of our modern ideas of the structure of the atom. A simplified diagram of an atom is shown in Figure 2.1.

Charge and mass of protons, neutrons and electrons

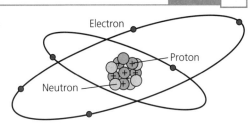

Figure 2.1 Structure of the atom

Atoms have a diameter of the order of 10^{-8} m, and consist of two parts:
- A central heavy nucleus, with diameter of the order of 10^{-15} m that contains:
  - **protons** — particles with a unit positive charge of $+1.6 \times 10^{-19}$ C and a mass (m_p) of 1.67×10^{-27} kg
  - **neutrons** — neutral particles with a mass slightly greater than that of a proton; the neutron mass (m_N) is 1.675×10^{-27} kg; $m_N = 1.0014\,m_p$
- **Electrons** orbiting the nucleus. These are particles with a negative charge of -1.6×10^{-19} C, equal and opposite to that of a proton. The mass of the electron (m_e) is 9.11×10^{-31} kg, or about 1/1836 of that of a proton. The number of these electrons is equal to the number of protons in a non-ionised atom

Note: all figures quoted are for the rest masses of the particles. Relativistic effects will be ignored.

> **Exam tip**
>
> Remember to use the correct SI units.

Specific charge of nuclei and of ions

A useful quantity is the **specific charge** of a particle. This is defined as follows:

$$\text{specific charge of a particle} = \frac{Q}{m}$$

where Q is the charge on the particle and m is its mass. The units for specific charge are coulombs per kg.

The specific charges of a number of particles are given in Table 2.1.

Table 2.1 Particles and their specific charges

Particle	Specific charge
A proton	$\dfrac{+1.6 \times 10^{-19}}{1.67 \times 10^{-27}} = +9.58 \times 10^{7}\,\text{C kg}^{-1}$
An electron	$\dfrac{-1.6 \times 10^{-19}}{9.11 \times 10^{-31}} = -1.76 \times 10^{11}\,\text{C kg}^{-1}$
A nucleus of carbon-12	$\dfrac{+(12 \times 1.6 \times 10^{-19})}{1.992 \times 10^{-27}} = +9.64 \times 10^{8}\,\text{C kg}^{-1}$

Proton and nucleon number

The **nucleon number** varies from 1 for the simplest form of hydrogen to about 250 for the heaviest elements. The **proton number** varies from 1 to just over 100 for the same range of particles.

> The **proton number** (Z) is the number of protons in a nucleus.
>
> The **nucleon number** (A) is the total number of neutrons and protons in a nucleus.

Nuclear notation

The correct way of writing down the structure of a nuclide, showing the proton and nucleon numbers, is shown in Figure 2.2.

Isotopes

Neon has 10 protons in its nucleus but may occur in a number of different forms with nucleon numbers of 20, 21 and 22, corresponding respectively to 10, 11 and 12 neutrons in the nucleus. These different forms are known as **isotopes** of neon.

> **Isotopes** are different atoms of the same chemical element, i.e. they have the same proton numbers, with different nucleon numbers.

Figure 2.3 shows the three isotopes of hydrogen.

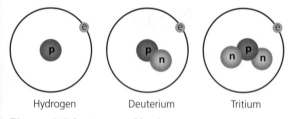

Figure 2.3 Isotopes of hydrogen

- The chemical properties of isotopes of the same element are identical.
- Their nuclear properties will be different and some of their physical properties, such as boiling point, are different as well

Exam tip

The nucleon number was previously called the mass number, while the proton number was known as the atomic number.

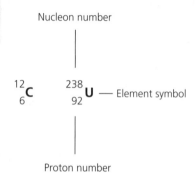

Nucleon number

$^{12}_{6}\text{C}$ $\quad$ $^{238}_{92}\text{U}$ — Element symbol

Proton number

Figure 2.2 Nuclide notation

Typical mistake

Confusing the nucleon and proton number in the nuclide notation.

Revision activity

Make a table showing the nuclear structure of two isotopes of four different elements.

Now test yourself

1 The proton number of uranium is 92. How many neutrons are there in the following two isotopes of uranium:
   (a) uranium-235 (nucleon number 235)
   (b) uranium-238 (nucleon number 238)
2 What is the nucleon number of the nucleus containing 26 protons and 28 neutrons? (This is an isotope of iron.)

Answers on p. 216

Stable and unstable nuclei

The strong nuclear force

REVISED

There are two kinds of particle in the nucleus of an atom — protons, carrying a unit positive charge, and neutrons, which are uncharged. The electrostatic repulsion between all those positively charged protons would tend to blow it apart were it not for the existence of another attractive force between the nucleons. This is known as the **strong nuclear force**.

> The **strong nuclear force** acts between particles in the nucleus and is responsible for the stability of the nucleus.

The strong nuclear force between two nucleons is a short-range force. It is attractive and acts up to a nucleon separation of about 3 fm (3×10^{-15} m). This 'holds the nucleons together' in the nucleus. However, at very small nucleon separations of less than 0.5 fm it becomes repulsive. The repulsive nature at these very small distances keeps the nucleons at a minimum separation.

In small nuclei the strong force from all the nucleons reaches most of the others in the nucleus but for nuclei with more protons and neutrons the balance becomes much finer. The nucleons are not held together so tightly and this can make the nucleus unstable.

Exam tip

The strong nuclear force is very short range, while the electrostatic force affects all the nuclei in the nucleus.

Alpha and beta emission

REVISED

Radioactive decay, the result of instability in a nucleus, is the emission of particles from the nucleus or a loss of energy from it as electromagnetic radiation.

The two types of particle emitted are the alpha particle (two protons and two neutrons — a helium nucleus) and the beta particle (an energetic electron).

Figure 2.4 shows the changes in nucleon and proton number due to the emission of either an alpha particle or a beta particle.

Example emissions of an alpha and a beta particle, expressed in equation form, are given below,

(a) Alpha emission:

$$^{226}_{88}\text{Ra} \rightarrow\, ^{222}_{86}\text{Rn} +\, ^{4}_{2}\alpha$$

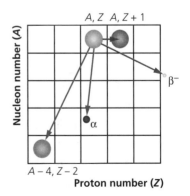

Figure 2.4 Alpha and beta emission

(b) Beta emission:

$$^{90}_{38}\text{Sr} \rightarrow {}^{90}_{39}\text{Y} + {}^{0}_{-1}\beta + \bar{\nu}_e$$

Note: the particle $\bar{\nu}_e$ is an **antineutrino**. The existence of **neutrinos** and antineutrinos is necessary to satisfy nuclear conservation laws (p. 22). They are both neutral particles with virtually no mass.

Knowing the true nature of alpha and beta particles, the equations can also be rewritten as:

(a) Alpha emission:

$$^{226}_{88}\text{Ra} \rightarrow {}^{222}_{86}\text{Rn} + {}^{4}_{2}\text{He}$$

(b) Beta emission:

$$^{90}_{38}\text{Sr} \rightarrow {}^{90}_{39}\text{Y} + {}^{0}_{-1}\text{e} + \bar{\nu}_e$$

(c) Beta-plus (positron) emission with the electron neutrino:

$$^{30}_{15}\text{P} \rightarrow {}^{30}_{14}\text{Si} + {}^{0}_{+1}\text{e} + \nu_e$$

> **Exam tip**
>
> Beta emission is the result of neutron decay within the nucleus.

Example

Plutonium-239 decays to form uranium-235.
(a) Is this by alpha or beta emission?
(b) Write down the nuclear equation to show this decay.

Answer

(a) alpha emission
(b) $^{239}_{94}\text{Pu} \rightarrow {}^{235}_{92}\text{U} + {}^{4}_{2}\text{He}$

Now test yourself

TESTED

3 Carbon-14 decays by beta emission.
  (a) What is the resulting nuclide?
  (b) Write down the full nuclear equation for this process.
4 Uranium-238 decays to form thorium-234.
  (a) Is this by alpha or beta decay?
  (b) Write down the full nuclear equation for this process.

Answers on p. 216

Particles, antiparticles and photons

Antiparticles

REVISED

All particles of matter have a corresponding antiparticle. The first antiparticle to be identified was the anti-electron or **positron**. The mass of the positron is the same as that of an electron (0.51 MeV — p. 21).

Protons, neutrons and neutrinos each have their antiparticle — the antiproton, the antineutron and the antineutrino. The masses of all these antiparticles are the same as those of their corresponding particles.

There is a whole set of antiparticles that 'mirror' the particles that make up our universe. These antiparticles would combine to form a 'new' type of matter known as **antimatter**.

> **Exam tip**
>
> Particles and antiparticles have opposite charges where this is appropriate (proton and electron).

Photon model of electromagnetic radiation

All objects at a temperature above absolute zero emit a range of wavelengths but the peak of the energy–radiation curve moves towards the short-wavelength, high-frequency, end as the temperature of the object is increased. For example, a hot piece of metal glows first red, then orange then yellow and finally white as its temperature is increased.

More and more energy is emitted as short-wave radiation (Figure 2.5).

The two curves represent objects at two different temperatures. The lower curve is the lower temperature.

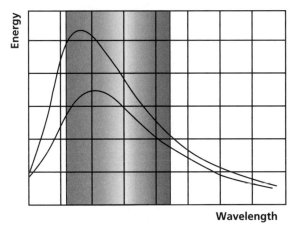

Figure 2.5 Energy–wavelength graph

In 1900 the problem of the energy distribution that had been puzzling scientists for some time was solved by Max Planck. He proposed that radiation was emitted not in a continuous stream of energy but in bundles of energy that we now call **photons**.

The energy of a **photon** is given by the formula:

photon energy (E) = hf

where f is frequency and h is the Planck constant, with a value of 6.63 × 10^{-34} J s.

Radiation of a higher frequency, and therefore a shorter wavelength, will be composed of photons that have a greater energy.

> **Typical mistake**
>
> Assuming that if an object, such as a lump of metal, does not glow it is not hot.

We can use this idea to calculate the number of photons emitted by a 100 W yellow light per second. (frequency of yellow light = 5×10^{14} Hz)

Answer

energy emitted by the light bulb every second = 100 J

energy of each quantum = $hf = 6.63 \times 10^{-34} \times 5 \times 10^{14} = 3.31 \times 10^{-19}$ J

Therefore:

number emitted per second per second = $\dfrac{100}{3.31 \times 10^{-19}} = 3.0 \times 10^{20}$ photons

The energy of each photon must be very small otherwise they would hurt when they hit you!

Now test yourself

TESTED

5 Calculate the energies of a photon of the following wavelengths:
   (a) gamma rays   wavelength 10^{-3} nm
   (b) X-rays   wavelength 0.1 nm
   (c) violet light   wavelength 420 nm
   (d) yellow light   wavelength 600 nm
   (e) red light   wavelength 700 nm
   (f) microwaves   wavelength 2 cm
   (g) radio waves   wavelength 254 m

Answer on p. 216

Particle annihilation

REVISED

When a positron meets an electron the two particles annihilate each other, converting their mass back into energy in the form of electromagnetic radiation (see Figure 2.6). Two gamma rays are needed to conserve momentum. The energy produced in this case is about 1.02 MeV.

A similar event will occur between any particle and its antiparticle.

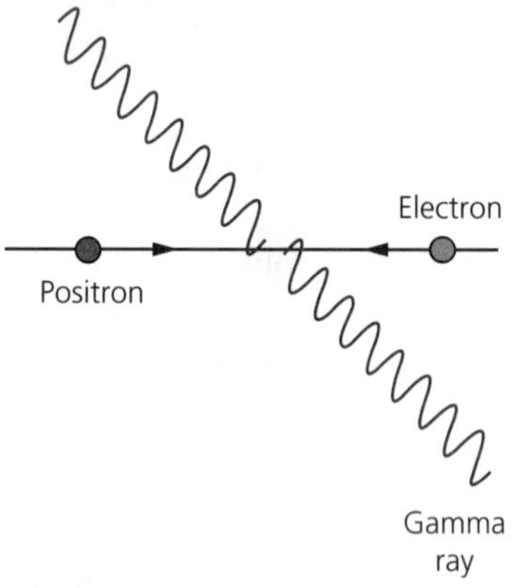

Figure 2.6 Particle–antiparticle annihilation

A proton collides with an antiproton and they annihilate each other. Calculate the energy released in MeV. (rest mass of a proton = 938 MeV; rest mass of an antiproton = 938 MeV)

Answer

energy released = $2 \times 938 = 1876$ MeV

In high-energy collisions between protons and antiprotons, for example those in the Large Hadron Collider at CERN, the particles annihilate each other. The sum of their mass energy and kinetic energy is converted into radiation and other particles. The initial energy of the proton and antiproton can be as high as 800 GeV (800×10^3 MeV).

Exam practice answers and quick quizzes at **www.hoddereducation.co.uk/myrevisionnotes**

Pair production

The reverse of particle annihilation can occur. When a gamma ray passes close to a nucleus it can interact with that nucleus, forming a positron and an electron. This is known as **pair production**. Matter and antimatter have been produced from energy (the gamma ray — Figure 2.7).

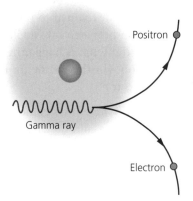

Figure 2.7 Pair production

Particle interactions

There are at present only four certainly known types of force, and these are listed below. The relative importance of each force in an interaction depends on the type of interaction being considered.

- The **gravitational force** acts between all particles with mass and is responsible for holding planets in orbit around the Sun. Range: infinite, varying as $1/d^2$.
- The **electromagnetic force** acts between all charged particles, and is the binding force of atoms and molecules. Range: infinite, varying as $1/d^2$
- The **weak force** is responsible for radioactive decay and the change in quark flavour. It acts between all particles. It is seen in lepton reactions such as the reaction between a neutrino and a muon. Range: about $10^{-3}\,fm$ ($10^{-18}\,m$).
- The **strong force** holds neutrons and protons together in a nucleus. It only acts between hadrons since they contain quarks. Range: about $3\,fm$. Repulsive up to $0.5\,fm$ and attractive from $0.5\,fm$ to $3\,fm$.

Exchange particles

These fundamental forces can be explained by describing them in terms of **exchange particles**. These are particles that are passed between the two interacting particles and so 'carry' the force between them. These exchange particles are shown in Table 2.2.

Table 2.2 Exchange particles

Force (interaction)	Particle name	Charge
Electromagnetic	Photon	0
Strong	Gluon	0
Gravitational	Graviton	0
Weak	W⁺	+e
	W⁻	−e
	Z	0

When an electron repels another electron they *both* emit a photon. These photons are 'exchanged' between the two electrons and this 'carries' the force to 'push them apart'. In the weak interaction that governs β⁻ and β⁺ decay, electron–proton collisions and electron capture, the exchange particles are the W⁻, the W⁺ and the Z respectively.

Exam tip

At the time of writing (2017) the exchange particle for gravitational force, the graviton, has not been discovered.

Exam tip

Remember that in the interaction *both* particles emit a particle or a photon, hence the name — exchange particles.

Typical mistake

Confusing exchange particles with 'actual' particles.

Particle interaction diagrams (Feynman diagrams)

These were developed to provide a clear method of showing the interaction between sub-nuclear particles. They are a way of representing what is happening between the two particles *during* an interaction. (In the following Feynman diagrams time goes from bottom to top.)

Each point where lines come together is called a vertex. At each vertex charge, baryon number and lepton number must be conserved. (For an explanation of these terms see p. 21.)

Electromagnetic force interaction

Figure 2.8 shows the interaction between two electrons. In classical physics the electrons, both with a negative charge, would repel each other. The diagram shows that this repulsion occurs because of the interchange of photons. Each electron emits a photon, which is then absorbed by the other electron. The photons in the interaction are known as **virtual photons** because they are emitted and absorbed in a time so short that the uncertainty principle is not violated. (To simplify the diagram only one of the virtual photons is shown.)

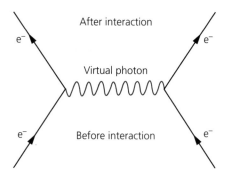

Figure 2.8 Electron–electron interaction

Weak force interaction

The weak force interaction in Figure 2.9 shows the emission of a β^- in the decay of a free neutron to a proton and an antineutrino. The exchange particle in this interaction is a W^-.

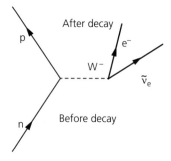

Figure 2.9 Neutron decay

The weak force interaction in Figure 2.10 shows the emission of a β^+ in the decay of a proton to a neutron and an antineutrino. The exchange particle in this interaction is a W^+.

Electron capture

Electron capture is a process in which an electron in the inner shell of an atom is absorbed by the nucleus, changing a nuclear proton to a neutron and simultaneously emitting a neutrino (Figure 2.11).

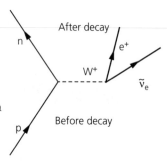

Figure 2.10 Proton decay

Electron–proton collision

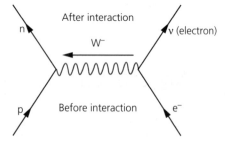

Figure 2.12 Electron–proton collision

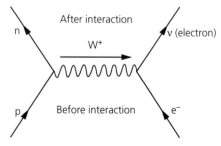

Figure 2.11 Electron capture

Exam practice answers and quick quizzes at **www.hoddereducation.co.uk/myrevisionnotes**

Classification of particles

Fundamental particles

REVISED

The fundamental particles that make up our universe can be divided into three main categories:

- hadrons — these particles 'feel' the strong nuclear interaction (force)
- leptons — these particles 'feel' the weak nuclear interaction
- exchange particles

These three categories are further divided as follows:

- **Hadrons** are composed of two or three quarks:
  - Baryons — protons, neutrons (each composed of three quarks)
  - Mesons — pions, kaons (each composed of two quarks)
  (Every particle has its own antiparticle.)
- **Leptons** — electrons, muons, neutrinos — subject to the weak interaction
  (Every particle has its own antiparticle.)
- **Exchange particles** — the interactions between particles in the previous two groups are carried by exchange particles (photons, gluons, $W^{\pm}$, Z and gravitons). For more on exchange particles, see p. 19

> **Typical mistake**
>
> Not remembering to which category a particle belongs.

Leptons and hadrons are summarised in Table 2.3.

Table 2.3 Leptons and hadrons

Particle	Symbol	Rest energy/ MeV	Charge/e	Lifetime
Leptons				
Electron	e	0.511	−1	$>4.6 \times 10^{26}$ years
Neutrino	v_e and v_μ	Very small	0	Stable
Muon	μ	105.7	−1	2.2×10^{-6} s
Hadrons				
Mesons				
Pion	π^+	139.6	+1	2.6×10^{-8} s
	π^0	135.0	0	0.8×10^{-16} s
Kaon	K^-	495	−1	1.24×10^{-8} s
Baryons				
Proton	p	938.3	+1	$>1 \times 10^{29}$ years
Neutron (free)	n	939.6	0	650 s

Lepton number

All leptons have a 'property' called a **lepton number**, L_e (= 1) for electrons and electron-neutrinos and L_μ (= 1) for muons and muon-neutrinos. Their antiparticles have lepton numbers of −1 for both positrons and antineutrinos.

Baryon number

All baryons and antibaryons are given a **baryon number**. This is +1 for baryons and −1 for anti-baryons.

To account for the strange behaviour of some hadrons a 'new' property of hadrons was proposed, called **strangeness** (S). Protons and neutrons have a strangeness of zero, while kaons have a strangeness of +1 or −1.

Exam tip

Remember that although mesons are hadrons they are not baryons and so their baryon number is 0.

Particle	S	Particle	S	Particle	S
n	0	e^+ and e^-	0	π	0
p	0	ν	0	K^+	+1

Conservation in particle interactions

REVISED

- The total charge is always conserved in a particle interaction.
- The total baryon number is always conserved in a particle interaction.
- The total lepton number is always conserved in a particle interaction.
- Strangeness is conserved in strong interactions but not in weak interactions

Examples

1 Show that the quantum number conservation laws for charge and lepton number are obeyed in the following reaction:

$n \rightarrow p^+ + e^- + \bar{\nu}_e$

Answer

	n	$\rightarrow$	p^+	+	e^-	+	$\bar{\nu}_e$
Charge	0		+1		−1		0
Lepton number (L_e)	0		0		1		−1
Strangeness (S)	0		0		0		0

Therefore the conservations laws are followed and the interaction will take place.

2 Show that the following reaction does not obey all the quantum number conservation laws and will therefore not happen:

$n + K^+ \rightarrow \pi^0 + \pi^+$

Answer

	n	+	K^+	$\rightarrow$	π^0	+	π^+
Charge	0		+1		0		+1
Baryon number	1		0		0		0
Strangeness (S)	0		+1		0		0

Therefore the conservations laws are not followed and the interaction will not take place.

Now test yourself

TESTED

6 Show that the quantum number conservation laws for charge and lepton number are obeyed in the following reaction:

$\mu^+ \rightarrow e^+ + \bar{\nu}_\mu + \bar{\nu}_e$

Answer on p. 216

Quarks and antiquarks

All hadrons are composed of particles called **quarks**. These were finally discovered in 1975 by the bombardment of protons by very high-energy electrons.

Exam tip

Leptons do not contain quarks. They themselves are considered to be fundamental particles.

At the present time (2017) quarks are thought to be the fundamental particles of matter. Quarks have fractional electric charge compared with the charge on the electron of −e.

The existence of quarks was confirmed by high-energy electron scattering from the nucleons. There are actually six quarks and their antiquarks, but we will only consider three types here (together with their antiquarks):

- the up quark (u)
- the down quark (d)
- strange quark (s)

See Figure 2.13.

Figure 2.13 Quarks

Note: the full list of quarks is up, down, strange, charm, bottom and top.

Properties of quarks

REVISED

Table 2.4 shows the properties of up, down and strange quarks.

Table 2.4 Quark properties

Quark	Symbol	Charge	Baryon number	Strangeness
Up	u	$+\frac{2}{3}$	$\frac{1}{3}$	0
Down	d	$-\frac{1}{3}$	$\frac{1}{3}$	0
Strange	s	$-\frac{1}{3}$	$\frac{1}{3}$	−1

Combinations of quarks

REVISED

Note: the colours of the quarks in the following diagrams are simply to make them distinguishable.

Baryons

Baryons are formed from combinations of three quarks or antiquarks (Figure 2.14).

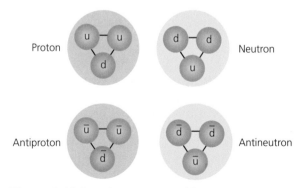

Figure 2.14 Quark structure of baryons

Mesons

Mesons are formed from combinations of two quarks or antiquarks (Figure 2.15).

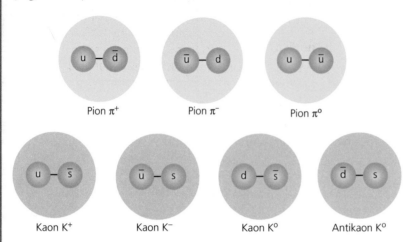

Figure 2.15 **Quark structure of mesons**

Applications of conservation laws

Quark model of beta emission

REVISED

The quark nature of the proton and neutron can be used to explain beta emission:

β^+ emission: $p \rightarrow n + \beta^+ + \nu$

Quark version:

β^+ emission (proton decay): $uud \rightarrow ddu + 0 + 0$

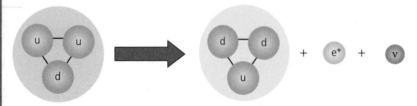

Figure 2.16 **Quark model of β^+ emission**

In proton decay an up quark changes into a down quark.

β^- emission: $n \rightarrow p + \beta^- + \bar{\nu}$

Quark version:

β^- emission (neutron decay): $ddu \rightarrow uud + 0 + 0$

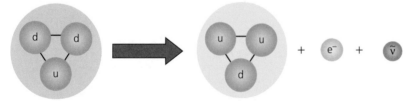

Figure 2.17 **Quark model of β^- emission**

When a neutron decays by β^- emission a down quark changes into an up quark.

Exam practice answers and quick quizzes at **www.hoddereducation.co.uk/myrevisionnotes**

The photoelectric effect

Photon explanation of threshold frequency

Figure 2.18 shows a charged clean zinc plate fitted to the top of a gold leaf electroscope. The plate may be positive or negative and various forms of radiation can be shone on it.

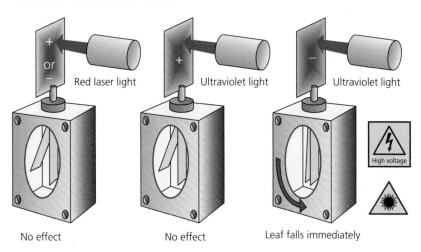

Red laser light   Ultraviolet light   Ultraviolet light

High voltage

No effect   No effect   Leaf falls immediately

Figure 2.18 The photoelectric effect

If the plate is positively charged no radiation has any effect. However, if the plate is given a negative charge to start with there *is* a difference. Using the laser-emitting red light has no effect, but when ultraviolet light is shone on the plate the electroscope is discharged and the leaf falls immediately. No effect can be produced with radiation of longer wavelength (lower frequency and smaller energy) no matter how long the radiation is shone on the plate.

When the ultraviolet radiation fall on the plate:
- no electrons are emitted from the plate if it is positive
- the number of electrons emitted per second depends on the intensity of the incident radiation
- the energy of the electrons depends on the frequency of the incident radiation
- there is a minimum frequency (f_0) below which no electrons are emitted no matter how long radiation fell on the surface

These results show that:
- The **threshold frequency** is the minimum frequency (f_0) that will cause electron emission from a given material. Photons with a lower frequency will never cause electron emission.
- The free electrons are held in the metal in a 'hole' in the electric field; this is called a **potential well**. Energy has to be supplied to them to enable them to escape from the surface (Figure 2.19).

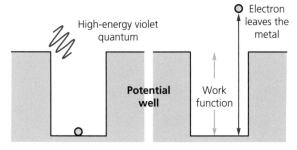

High-energy violet quantum

Electron leaves the metal

Potential well

Work function

Figure 2.19 The potential well

If radiation with a frequency above that of the threshold frequency is shone on a metal plate electrons are emitted spontaneously. One quantum of radiation (a photon) of a high enough frequency has enough energy to 'kick the electron out' in one go.

The photoelectric effect is very good evidence for the particle nature of electromagnetic waves.

The amount of energy needed to just release a photoelectron is known as the **work function** (φ) for the metal. This can be expressed in terms of the threshold frequency (f_0).

> **work function (φ) = hf_0**
>
> where h is the Planck constant (6.63×10^{-34} Js)

Exam tip

Remember to use SI units for the work function in calculations.

Example

If the work function of silver is 7.6×10^{-19} J, calculate the threshold frequency for a clean silver surface.

Answer

$$\text{threshold frequency } (f_0) = \frac{W}{h} = \frac{7.6 \times 10^{-19}}{6.63 \times 10^{-34}} = 1.15 \times 10^{15} \text{ Hz}$$

(This is in the ultraviolet region of the spectrum.)

Einstein's photoelectric equation

REVISED

If a quantum of radiation with an energy (hf) greater than the work function φ, and therefore a frequency greater than f_0, falls on a surface an electron will escape from the surface and be emitted with some residual kinetic energy (E_k).

The energy of the incident quantum (hf) is the sum of the work function of the metal ($\varphi = hf_0$) and the maximum kinetic energy of the electron (E_k) (Figure 2.20). This is expressed by Einstein's photoelectric equation:

$$hf = \varphi + E_k = hf_0 + E_k$$

Stopping potential

REVISED

If we put a collecting electrode in front of the emitting surface in a vacuum we can detect the photoelectrons as a small current. If the collecting electrode is made slightly negative compared with the emitting surface the electrons will find it difficult to get to it and electrons will only do that if their energy is greater than the 'height' of the potential barrier.

Electrons will be only detected if $E_k > eV$ where V is the potential difference between the plate and the emitting surface.

If V is increased so that no more electrons can reach the detector, this value for the potential is called the **stopping potential** for that surface and radiation.

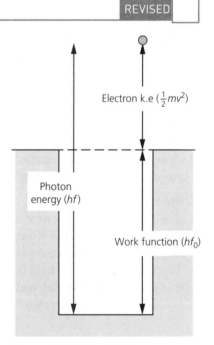

Electron k.e ($\frac{1}{2}mv^2$)

Photon energy (hf)

Work function (hf_0)

Figure 2.20 Einstein's photoelectric equation

> **Example**
>
> Calculate the maximum kinetic energy of an electron that is emitted from a magnesium surface when light of wavelength 180 nm falls on it. (work function for magnesium = 5.9×10^{-19} J)
>
> **Answer**
>
> frequency of incident radiation $= \dfrac{3 \times 10^8}{180 \times 10^{-9}} = 1.67 \times 10^{15}$
>
> kinetic energy $= hf - hf_0 = hf - \varphi$
>
> $= (6.63 \times 10^{-34} \times 1.67 \times 10^{15}) - 5.9 \times 10^{-19} = 1.11 \times 10^{-18} - 5.9 \times 10^{-19} = 5.17 \times 10^{-19}$ J

Now test yourself

TESTED ☐

7 Light with a wavelength of 150 nm is needed to cause photoelectric emission from the surface of a piece of metal. Calculate the work function for that metal.

8 Radiation of wavelength 120 nm falls on a zinc plate. Electrons are emitted with a maximum energy of 6.6×10^{-19} J.
   Calculate:
   (a) the energy of a quantum of the incident radiation
   (b) the work function for zinc
   (c) the threshold frequency for zinc
   (speed of light, $c = 3 \times 10^8$ m s^{-1}; Planck constant, $h = 6.63 \times 10^{-34}$ J s)

Answers on p. 216

> **Exam tip**
>
> Do not forget to convert nm to m when using the photoelectric equation.

> **Revision activity**
>
> Summarise the main features of photoelectric emission in a mind map or table and emphasise what it shows about the nature of radiation.

Collisions of electrons with atoms

Ionisation and excitation

REVISED ☐

A simplified version of an energy level diagram for electrons in an atom is shown in Figure 2.21. The electrons are spread through the energy levels. No electron can have an energy state between the levels.

In hydrogen there is just one orbiting electron. The electron is usually in its unexcited or **ground state** — level 1 (Figure 2.22(a)).

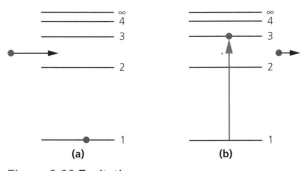

Figure 2.22 Excitation

If energy is put into the atom in the form of radiant energy or by an inelastic collision with a charged particle, an electron is raised to a higher energy level and is said to be **excited** and in an **excited state**.

In Figure 2.22(b) the electron has been raised to level 3 and the colliding electron has lost some energy.

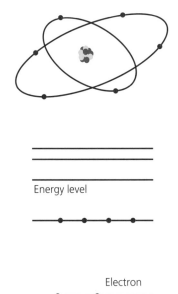

Figure 2.21 Electrons in an atom

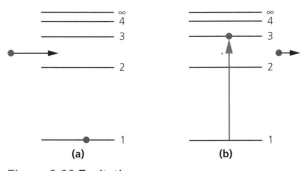

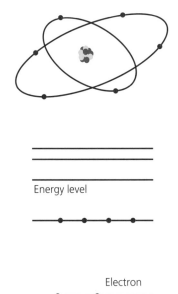

AQA A-level Physics 27

If the collision with an incoming electron is sufficiently violent an electron within the atom can be given enough energy to raise it to the level marked with an infinity symbol. This level is called the **ionisation level**.

If the energy input is great enough to raise it above that level the electron will escape from the atom altogether. This is called **ionisation** (Figure 2.23).

The removal of one (or more) electrons will leave the atom with a net positive charge — it has become a positive ion.

The energy required to ionise a hydrogen atom is 21.8×10^{-19} J. This assumes that the electron starts off in its ground state.

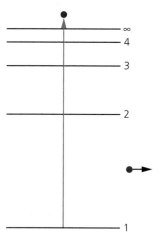

Figure 2.23 Ionisation

The fluorescent lamp

The fluorescent lamp is a sealed glass tube containing a gas such as mercury at low pressure. A filament in the tube produces electrons by thermionic emission and these move at speed through the tube due to a large electric field between the two electrodes.

Collisions between these electrons and the mercury atoms excite atoms of the gas. Transitions within the mercury atoms give out radiation, which is mostly in the ultraviolet region of the spectrum, and finally these ultraviolet photons interact with the phosphor on the glass walls of the tube, producing visible light.

The electron volt

REVISED

Joules are very large units when subatomic particles are considered. A much smaller unit known as the **electron volt** is used when stating their energies.

> An **electron volt** (eV) is the energy gained by an electron when it is accelerated through a potential difference of 1 volt.
>
> $$1\,eV = 1\,V \times 1.6 \times 10^{-19}\,C = 1.6 \times 10^{-19}\,J$$

Yellow light, wavelength 600 nm, has a frequency of 5×10^{14} Hz and so the energy of a photon of yellow light is:

energy $= hf = 6.63 \times 10^{-34} \times 5 \times 10^{14} = 3.31 \times 10^{-19}$ J

which, when expressed in electron volts, is:

$$\frac{3.31 \times 10^{-19}}{1.6 \times 10^{-19}} = 2.07\,eV$$

Larger energies can be expressed in keV (10^3 eV) and MeV (10^6 eV).

Example

Calculate the energy of a photon of ultraviolet light with a wavelength of 100 nm in both (a) joules and (b) electron volts. (Planck constant, $h = 6.63 \times 10^{-34}$ J s; speed of light, $c = 3 \times 10^8$ m s^{-1}; 1 eV $= 1.6 \times 10^{-19}$ J)

Answer

frequency, $f = \dfrac{c}{\lambda} = \dfrac{3 \times 10^8}{100 \times 10^{-9}} = 3 \times 10^{15}$ Hz

(a) $E = hf = 6.63 \times 10^{-34} \times 3 \times 10^{15} = 1.99 \times 10^{-18}$ J

(b) $E = \dfrac{1.99 \times 10^{-18}}{1.6 \times 10^{-19}} = 12.4\,eV$

Energy levels and photon emission

Atomic line spectra

When light from an incandescent monatomic gas is viewed with a spectroscope a spectrum similar to the one shown on Figure 2.24 is seen.

Figure 2.24 Atomic line spectrum

The spectrum shows a series of bright lines, which is very good evidence for the structure of the atom. The simplest spectrum is that of hydrogen. When an electron drops from one level to another a quantum of radiant energy known as a photon is emitted and this gives a line in the hydrogen spectrum.

The greater the energy transition the higher the frequency of the emitted radiation. The separation of the energy levels in the atom can be predicted from the wavelengths, and hence frequencies, of the radiation emitted (Figure 2.25).

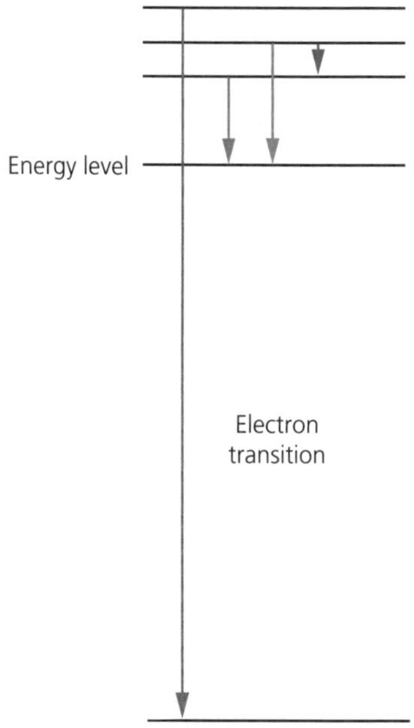

Energy level

Electron transition

Figure 2.25 Electron transitions

The spectrum of atomic hydrogen

The energy levels in atomic hydrogen are shown in Figure 2.26.

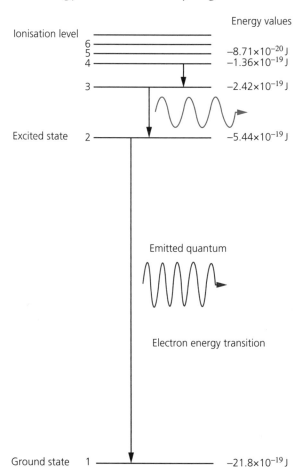

Figure 2.26 Energy levels in hydrogen

When an electron falls from one level to another energy is emitted in the form of a photon of radiation. The energy of this photon, and therefore its frequency and wavelength, is determined by the difference in energy between the two levels.

$$\text{photon energy} = hf$$

The frequency (f) of the photon emitted is related to the difference between the two levels ($E_1 - E_2$) by the equation:

$$hf = E_1 - E_2$$

Typical mistake

Forgetting to convert eV to joules when calculating the wavelength and/or frequency of emitted radiation in a transition.

Example

Calculate the frequency of a photon produced by an electron transition between level 4 and level 2 in the hydrogen atom. (Planck constant, $h = 6.63 \times 10^{-34}$ J s)

Answer

$E_1 - E_2 = (5.44 - 1.36) \times 10^{-19}$ J $= 4.08 \times 10^{-19}$ J

$$f = \frac{E_1 - E_2}{h} = \frac{4.08 \times 10^{-19}}{6.63 \times 10^{-34}} = 6.15 \times 10^{14} \text{ Hz}$$

Radiation of this frequency has a wavelength of $\frac{c}{f} = 488$ nm.

Typical mistake

Using eV and nm when finding the frequency and wavelength.

Now test yourself

TESTED

9 Calculate the frequency and wavelength of a photon of radiation emitted due to an electron transition in a hydrogen atom between level 2 and level 1 (use Figure 2.26).

Answer on p. 216

Wave–particle duality

Electron diffraction

REVISED

If electrons have wave properties they should show the characteristics of waves such as interference and diffraction. The fact that electron diffraction can be observed suggests that particles do have wave properties.

The diagrams in Figure 2.27 show the effects produced by electron diffraction through a thin graphite sheet.

(a) Low accelerating voltage   (b) High accelerating voltage

Figure 2.27 Electron diffraction rings

Electron diffraction is very good evidence for the wave nature of particles.

If the accelerating voltage is increased the energy and momentum of the electrons is increased, and the diameter of a given ring gets less, showing a smaller angle of diffraction.

The wave theory of particles suggests that this is because the electrons' wavelength has also decreased. This is exactly similar to the observation that blue light with a short wavelength and high energy is diffracted through a smaller angle than low-energy red light.

Revision activity

Make a simple mind map of electron diffraction, showing in which part of the experiment the electrons' wave properties predominate and where their particle properties predominate.

The de Broglie wavelength

REVISED

Louis de Broglie proposed that a particle of mass m travelling with a velocity v would have a wavelength λ given by the equation:

$$\text{wavelength, } \lambda = \frac{h}{mv}$$

where h is the Planck constant and mv is the momentum of the particle. The intensity of the wave at any point represents the probability of the particle being at that point.

An electron accelerated through a potential difference of V volts will gain electrical energy ($E = eV$) and hence kinetic energy $\frac{1}{2}mv^2$. The wavelength associated with the electron at that energy is given by:

$$\text{electron wavelength, } \lambda = \frac{12.27 \times 10^{-10}}{\sqrt{V}}$$

An electron of high energy has a smaller wavelength than one of low energy.

Example 1

Calculate the wavelength associated with an electron that has been accelerated through a potential difference of 5 kV.

Answer

$$\text{wavelength} = \frac{12.27 \times 10^{-10}}{\sqrt{V}} = \frac{12.27 \times 10^{-10}}{\sqrt{5000}}$$

$$= 1.74 \times 10^{-11}\,\text{m} = 0.017\,\text{nm}$$

(Compare this value with that for yellow light — about 600 nm.)

Strangely, whether a particle behaved like a particle or a wave seemed to be influenced by the nature of the experiment used.

Example 2

Calculate the wavelength of an electron emitted by a nucleus at 0.9 c. (mass of an electron travelling at this speed = $2.4 \times 9 \times 10^{-31}$ kg at this speed; $c = 3 \times 10^8\,\text{ms}^{-1}$)

Answer

$$\text{wavelength, } \lambda = \frac{6.63 \times 10^{-34}}{2.4 \times 9 \times 10^{-31} \times 3 \times 10^8} = 1.0 \times 10^{-12}\,\text{m}$$

Now test yourself

TESTED

10 Calculate the wavelength associated with a proton moving at $10^7\,m\,s^{-1}$. (mass of a proton = $1.67 \times 10^{-27}\,kg$; Planck constant = $6.63 \times 10^{-34}\,J\,s$)

Answer on p. 216

Exam practice

Use the following values where needed:

speed of electromagnetic radiation in free space, $c = 3 \times 10^8\,m\,s^{-1}$

Planck constant, $h = 6.63 \times 10^{-34}\,J\,s$

1 The following equation represents the alpha emission from a uranium nucleus:

$$^{235}_{92}U \rightarrow\, ^{a}_{b}Th + ^{4}_{2}\alpha$$

   (a) What are the numbers a and b? [2]
   (b) What do they represent? [2]

2 Which of the following combinations of alpha (α) and beta (β) particles can $^{214}_{84}Po$ emit and become another isotope of polonium?

   A α and 4β B α and 2β C α and β D 2α and β [1]

3 (a) What are the four properties conserved in particle interactions? [4]
   (b) Explain whether the following reaction obeys nuclear conservation laws. [2]

$$p + n \rightarrow p + \mu^+ + \mu^-$$

4 This question is about quarks.
   (a) How many quarks make up (i) a baryon, (ii) a meson? [2]
   (b) Write down the quark version of the decay of a neutron by beta-minus emission. [2]

5 The quark composition of an antiproton is:

   A uud   B ddu   C $\bar{u}\bar{u}\bar{d}$ D $u\bar{u}d$ [1]

6 'Photon' is the name given to:
   A a unit of energy
   B an electron emitted from a metal surface by incident radiation
   C a positively charged atomic particle
   D a quantum of electromagnetic radiation [1]

7 (a) (i) What is meant by the 'work function' in the photoelectric effect? [2]
       (ii) Which electrons are emitted in the photoelectric effect? [1]
   (b) Radiation of wavelength 180 nm ejects electrons from a potassium plate whose work function is 2.0 eV.
       (i) What is the maximum energy of the emitted electrons? [3]
       (ii) What is the maximum wavelength that will cause electron emission? [2]

8 An electron makes a transition from level 4 (energy $-0.85\,eV$) to level 3 (energy $-1.5\,eV$) in a hydrogen atom.
   (a) Calculate the wavelength of the radiation emitted. [2]
   (b) Suggest in which region of the electromagnetic spectrum this radiation lies. [1]

9 The wavelength of radiation emitted when an electron in an atom makes a transition from an energy state E_1 to one of energy E_2 is:

   A $\dfrac{hc}{E_2} - \dfrac{hc}{E_1}$

   B $\dfrac{E_1}{hc} - \dfrac{E_2}{hc}$

   C $\dfrac{hc}{E_1} - \dfrac{hc}{E_2}$

   D $\dfrac{hc}{E_2 - E_1}$ [1]

10 In an electron beam experiment the wavelength of an electron moving at $4.7 \times 10^6 \, \text{m s}^{-1}$ was found to be 0.155 nm.

(a) What value does this give for the rest mass of the electron? (At this speed relativistic effects can be ignored.) [2]

(b) The beam of electrons is now diffracted using a graphite sheet. What effect would a decrease of electron accelerating voltage have on the diameter of the diffraction rings? [1]

(c) What does the size of the diffraction rings and the wavelength of the electrons show about the spacing of the atoms in the graphite sheet? [2]

Answers and quick quiz 2 online

ONLINE

Summary

You should now have an understanding of:

- structure of the atom — atoms are composed of a nucleus of neutrons and protons with a cloud of electrons orbiting it
- stable and unstable nuclei — some nuclei are unstable and will lose energy by the emission of a particle (alpha and/or beta) or electromagnetic radiation (a gamma ray)
- particles, antiparticles and photons — all particles have their corresponding antiparticle; the Planck constant (h) is used to find the energy of a quantum of radiation
- particle interactions — there are four fundamental forces (gravity, electromagnetic, weak and strong), each carried by its own exchange particle
- classification of particles into hadrons (baryons and mesons) and leptons
- quarks and antiquarks — the fundamental 'building blocks' of hadrons; there are three types of quark — up, down and strange

- conservation laws — charge, baryon number, lepton number and strangeness are conserved in all nuclear interactions
- the photoelectric effect — the spontaneous emission of electrons from a surface due to incident radiation if its frequency is high enough
- collisions of electrons with atoms – ionisation occurs when an electron is removed from an atom
- an electron volt (eV) as a small unit of energy
- energy levels and photon emission — when an electron 'falls' from one energy level to another radiation is emitted; the frequency of this depends on the size of the energy transition
- wave–particle duality — particles can behave like waves and waves can behave like particles

3 Waves

Progressive waves

A progressive wave motion transmits energy from the source through a material or a vacuum without transferring matter. Wave motion can occur in many forms, such as water waves, sound waves, radio waves, light waves and mechanical waves.

Waves: basic properties REVISED

Waves are produced by the oscillation of particles or electric and magnetic fields. They are defined by the following set of basic properties:

- **Wavelength (λ)** is the distance between any two successive corresponding points on the wave, for example between two maxima or two minima.
- **Displacement (y)** is the distance from the mean, central, undisturbed position at any point on the wave.
- **Amplitude (a)** is the maximum displacement from zero to a crest or a trough.
- **Frequency (f)** is the number of vibrations per second made by the wave. Frequency is measured in hertz (Hz). A frequency of 1 Hz is a rate of vibration of one oscillation per second. High frequencies are measured in kilohertz (kHz) (1 kHz = 1000 Hz) and megahertz (MHz) (1 MHz = 1 000 000 Hz).
- **Period (T)** is the time taken for one complete oscillation ($T = 1/f$).
- **Phase (ε)** is a term related to the displacement at zero time (see p. 35).
- **Path difference** is the difference in distance travelled by two waves from their respective sources to a common point.
- **Speed (c)** is a measure of how quickly energy is transmitted from place to place by the wave motion.

wave speed (c) = frequency (f) × wavelength (λ)

> **Example**
>
> If a note played on a guitar has a frequency of 440 Hz, what is its wavelength? (speed of sound in air = 330 m s^{-1})
>
> Answer
>
> $$\text{wavelength} = \frac{\text{velocity}}{\text{frequency}} = 0.75\,\text{m}$$

> **Typical mistake**
>
> Taking a wavelength to be the distance along the wave from crest to trough and not crest to crest.

> **Typical mistake**
>
> Taking the amplitude to be the distance from a trough to a crest and not from the 'axis' to a crest.

Now test yourself TESTED

1 There are two radio stations broadcasting on the FM radio band. One has a frequency of 101.7 MHz and the other a frequency of 100 MHz. (speed of radio waves = 3 × 10^8 m s^{-1})
   What is the wavelength of the radio waves from the station with the shortest wavelength?

2 Explain why an astronaut on the surface of the Moon would be able to see a spacecraft descending towards the lunar surface but would not be able to hear the sound of the rocket engines. (Assume that they have a microphone fitted to the outside of their space helmet.)

Answers on p. 216

Phase and phase difference

The **phase** of a wave is related to the displacement of a specific point (say a crest) on the wave at zero time. The **phase difference** between two waves is the difference between the positions of the crests on the two waves.

When the positions of the crests and troughs of two waves coincide the waves are in phase. When the crests of one wave coincide with the troughs of the other the waves are out of phase (Figure 3.1). In this case the phase difference between the two waves is π radians, or 180°. Waves with a different phase difference would show a different shift along the time axis.

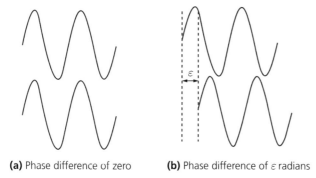

(a) Phase difference of zero   **(b)** Phase difference of ε radians

Figure 3.1 Phase and phase difference

Longitudinal and transverse waves

Types of wave

Wave motion occurs as one of two types: **longitudinal** and **transverse**.

Longitudinal waves

In a longitudinal wave (Figure 3.2) the oscillation is along the direction of propagation of the wave, for example sound waves and some mechanical waves.

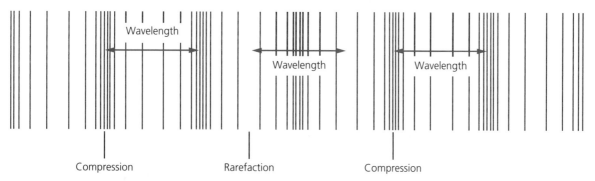

Figure 3.2 Longitudinal waves

In a longitudinal wave the particles of the material through which the wave is travelling move from side to side along the wave direction as the wave passes by. This oscillatory movement produces places of low pressure (**rarefaction**) and places of high pressure (**compression**). For this reason a longitudinal wave is sometimes called a pressure wave.

Transverse waves

In a transverse wave (Figure 3.3) the oscillations are at right angles to the direction of propagation of the wave, for example water waves, most electromagnetic waves and some mechanical waves.

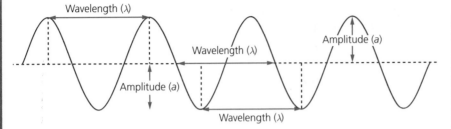

Figure 3.3 **Transverse waves**

Polarisation

REVISED

A wave in which the plane of vibration is constantly changing is called an **unpolarised** wave. When the vibrations of a transverse wave are in one plane only then the wave is said to be **polarised** (Figure 3.4).

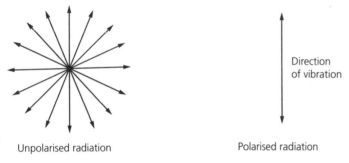

Unpolarised radiation

Polarised radiation

Direction of vibration

Figure 3.4 **Polarisation**

It is important to realise that transverse waves can be polarised while longitudinal waves cannot. Therefore if a set of waves can be polarised it is very good evidence that these waves are transverse. In Figure 3.5 it is clear that oscillations along the line of propagation will be unaffected by the polariser.

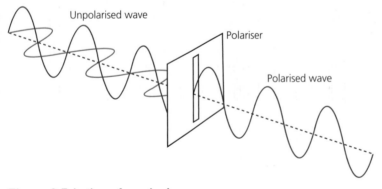

Unpolarised wave

Polariser

Polarised wave

Figure 3.5 **Action of a polariser**

All electromagnetic waves travel at the same speed in a vacuum.

Applications of polarisation

The uses of polarisation include polarising glasses for viewing 3D films, LCD displays, photographic filters, stress analysis investigation using transparent plastic specimens and 'Polaroid' sunglasses, which reduce the glare from reflected sunlight. This last effect is due to the polarisation of reflected light from a surface.

Polarisation is also important for the transmission and reception of TV signals. The transmitting aerial and the receiving aerial must be aligned in the same direction for optimum signal reception.

Now test yourself

TESTED

3 (a) Draw a diagram to show the relative alignment direction for rod-type TV transmitting and receiving aerials.
  (b) What would happen to the received signal as the receiving aerial was slowly rotated about an axis parallel to the direction of propagation of the incoming signal?

Answers on p. 216

Principle of superposition of waves and formation of stationary waves

Principle of superposition

REVISED

Unlike particles, waves can pass through each other when then overlap.

> The **principle of superposition** states that when two waves meet, the resulting displacement is the vector sum of the displacements due to each pulse at that point.

Formation of stationary waves

REVISED

A **stationary wave**, or standing wave, is one in which the amplitude varies from place to place along the wave. Figure 3.6 shows a stationary wave. The amplitude at point 1 is a_1, that at point 2 is a_2 and that at point 3 is a_3. The displacement (y) at these points varies with time.

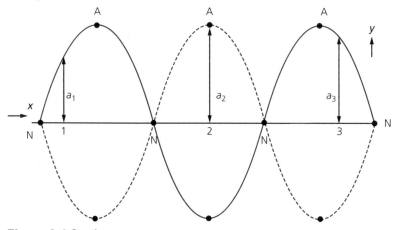

Figure 3.6 Stationary waves

Note that there are places where the amplitude is zero and, halfway between, places where the amplitude is a maximum; these are known as **nodes** (labelled N) and **antinodes** (labelled A) respectively.

Any stationary wave can be formed by the addition of two travelling waves moving in opposite directions.

A string is fixed between two points. If the centre of the string is plucked vibrations move out in opposite directions along the string. This causes a transverse wave to travel along the string. The pulses travel outwards along the string and when they reach each end of the string they are reflected.

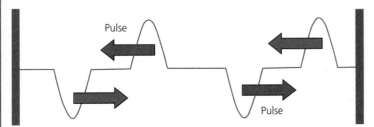

Figure 3.7 Pulses moving along a string

The two travelling waves then interfere with each other to produce a standing wave in the string. In the fundamental mode of vibration there are points of no vibration, or nodes, at each end of the string and a point of maximum vibration, or antinode, at the centre.

Notice that there is a phase change when the pulse reflects at each end of the string.

The frequency of the standing wave on a string depends on the length of the string (L), its tension (T) and the mass per unit length of the string (μ).

For the first harmonic the frequency is given by the formula:

$$\text{frequency}, f = \frac{1}{2L}\sqrt{\frac{T}{\mu}}$$

The first three harmonics for a vibrating string are shown in Figure 3.8.

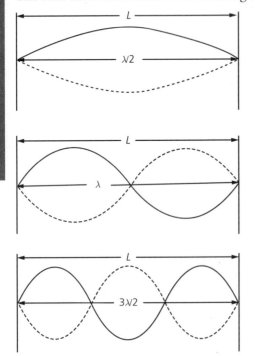

Figure 3.8 Nodes and antinodes on strings

Stationary waves can be formed on strings, as on a guitar or cello, and also using microwaves and sound.

Example 1

A stretched string is plucked at the centre and then lightly touched one quarter of the way from one end. Draw the resulting wave that is formed on the string.

Answer

When it is lightly touched a node will be produced at that point. The resulting waveform is shown in Figure 3.9.

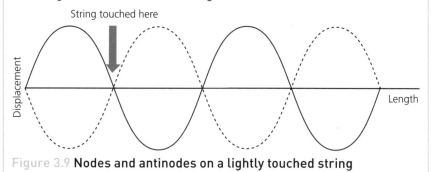

Figure 3.9 **Nodes and antinodes on a lightly touched string**

Example 2

A 70 cm long stretched string is plucked so that it vibrates in its second harmonic mode. If the tension in the string is 200 N and the mass per unit length is 1.2 g what note will be heard?

Answer

$$\text{frequency, } f = \frac{1}{L}\sqrt{\frac{T}{\mu}}$$

$$= (1/0.7)\sqrt{(200/1.2 \times 10^{-3})} = 583\,\text{Hz}$$

Note the use of the formula for the second harmonic.

Now test yourself

TESTED

4 What is the distance between adjacent nodes on a standing wave in terms of the wavelength of the standing wave?
5 Why must there always be nodes at the end of a standing wave on a stretched string?
6 A motorist drives along a motorway at a steady speed of $30\,\text{m s}^{-1}$ between two cities listening to the car radio. As she travels along she notices that the radio signal varies in strength, 5s elapsing between successive maxima. Explain this effect and calculate the wavelength of the radio signal to which she is tuned.

Answers on p. 216

Required practical 1

Standing waves on a stretched string

This can be demonstrated by fixing one end of a string to a vibration generator and passing the other end over a bench pulley with a weight fixed to the lower end. When the vibration generator is connected to a signal generator it will vibrate the string. Adjusting the tension and the length of the string and the driving frequency of the signal generator will give standing waves on the string.

Interference

Phase difference

When two waves meet at a point the resulting disturbance depends on the amplitudes of both waves at that point. This will depend on the **phase difference** between them. The formation of this disturbance is due to the superposition of the two waves and is called interference.

Coherent and incoherent sources

Two separate light sources, such as two light bulbs, cannot be used as sources for a static interference pattern because although they may be monochromatic the light from them is emitted in a random series of pulses. The phase difference that exists between one pair of pulses may well be quite different from that between the next pair of pulses (Figure 3.10).

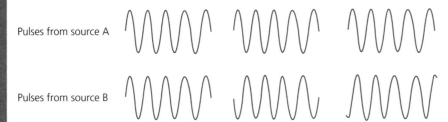

Pulses from source A

Pulses from source B

Figure 3.10 Coherent and incoherent sources

An interference pattern still occurs but it changes so rapidly that you get the impression of uniform illumination.

> Sources with synchronised phase changes between them are called **coherent sources** and those with random phase changes are called **incoherent sources**.

If the crest of one wave meets the crest of the other the waves are said to be in phase and the resulting intensity will be large. This is known as **constructive interference**. If the crest of one wave meets the trough of the other (and the waves are of equal amplitude) they are said to be out of phase by π and the resulting intensity will be zero. This is known as **destructive interference**.

There will be many intermediate conditions between these two extremes that will give a small variation in intensity.

This phase difference can be produced by allowing the two sets of waves to travel different distances. This difference in distance of travel is called the **path difference** between the two waves.

The diagrams in Figure 3.11 show two waves of equal amplitude with different phase and path differences between them. The first pair have a phase difference of zero and a path difference of a whole number of wavelengths, including zero. This gives constructive interference. The second pair have a phase difference of π or 180° and a path difference of an odd number of half-wavelengths, giving destructive interference.

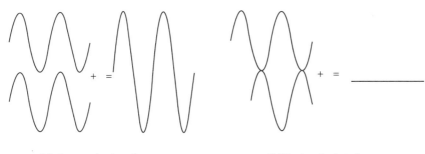

(a) Constructive interference   **(b)** Destructive interference

Figure 3.11 Constructive and destructive interference

To obtain a static interference pattern at a point (that is, one that is constant with time) we must have:
● two sources of the same wavelength, and
● two sources that have a constant phase difference between them.

This condition is met by two speakers connected to a signal generator because the sound waves that they emit are continuous — there are no breaks in the waves.

The laser — coherent light and safety

The problem of coherence of a source is overcome by using a laser. This emits a continuous beam of coherent light, with no abrupt phase changes (Figure 3.12).

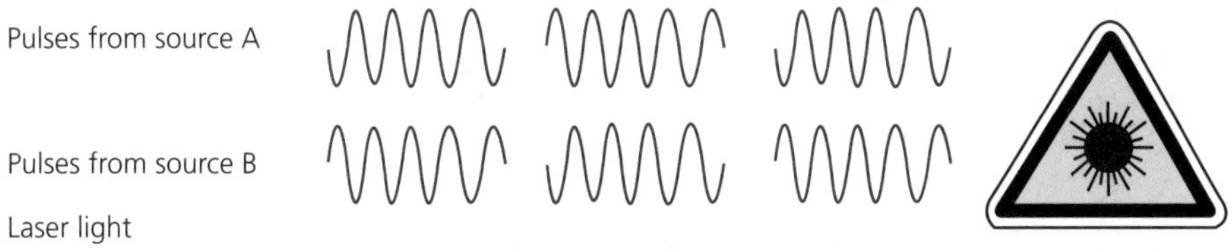

Pulses from source A

Pulses from source B

Laser light

Figure 3.12 Light waves emitted by a laser

Care must be taken when using lasers to avoid eye damage. The major problem with a laser is the power density. Light from a 100 W light bulb diverges and so the power density at a distance of 2 m from the source is $2\,\text{W}\,\text{m}^{-2}$. However a laser beam diverges very little. It is about 2 mm in diameter at a distance of 2 m from a 1 mW laser and so the power density here can be as high as $1.25 \times 10^4\,\text{W}\,\text{m}^{-2}$.

Two-source interference systems

A static interference pattern can be obtained using a single source and splitting the beam in two, as in the double-slit method. Light from a narrow single source (S) falls on two parallel slits (S_1 and S_2). This effectively gives two coherent sources since any phase changes in one source will also occur in the other (Figure 3.13).

Two sources, S_1 and S_2, emit waves of equal wavelength and these waves meet at a point P on a screen. In Figure 3.14(a) the path difference is 0, producing constructive interference. In Figure 3.14(b) it is half a wavelength and so destructive interference results.

The resulting interference pattern and the fringes formed are shown in Figure 3.15.

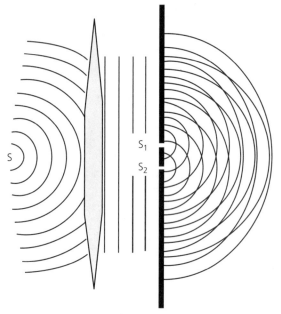

Figure 3.13 A double-slit interference system

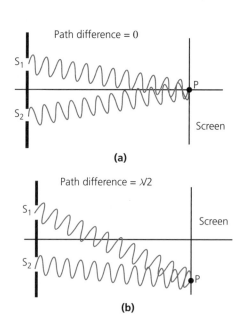

Figure 3.14 Double-slit interference

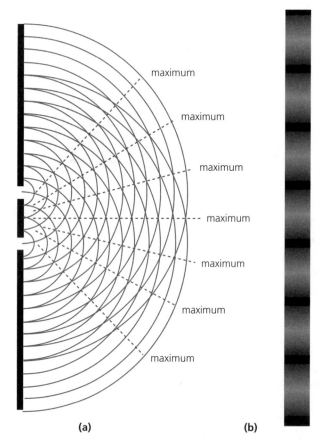

Figure 3.15 Interference pattern and fringes

The fringe spacing (or fringe width) for light of wavelength λ is given by the formula:

$$\text{fringe spacing, } w = \frac{\lambda D}{s}$$

where s is the slit spacing and D is the distance of the double slits from the screen.

Revision activity

Draw a full-size interference pattern with wavelength of 1 cm and plot the points of constructive and destructive interference.

Exam practice answers and quick quizzes at **www.hoddereducation.co.uk/myrevisionnotes**

Calculate the fringe spacing for light of wavelength 600 nm in a double-slit experiment where the double slits are separated by 0.8 mm and the screen is placed 75 cm from them.

Answer

fringe width $(w) = \dfrac{\lambda D}{d} = \dfrac{600 \times 10^{-9} \times 0.75}{0.80 \times 10^{-3}} = 5.6 \times 10^{-4}\,\text{m} = 0.56\,\text{mm}$

Now test yourself

7 Light of wavelength 600 nm falls on a pair of double slits that are 0.5 mm apart. Calculate:
(a) the fringe separation on a screen 90 cm away from the double slits
(b) the distance and direction that the screen has to be moved to get the same fringe separation with light of wavelength 500 nm
8 Calculate the wavelength of the light that will give an interference pattern with a fringe width of 4.5 mm on a screen 4 m from a pair of slits with a slit separation of 0.6 mm.

Answers on p. 216

Required practical 2(i)

Young's double-slit experiment

The interference of light can be observed using a blackened glass slide on which two fine slits have been engraved parallel to each other and less than a millimetre apart. When the slide is illuminated with monochromatic light a series of fringes will be seen on a screen placed on the other side of the slide from the source. Moving the screen away from the slits will increase the fringe width, as will the use of light of a longer wavelength.

Diffraction

Basic principles

When a wave hits an obstacle it does not simply go straight past, but bends round the obstacle. The same type of effect occurs at a hole — the waves spread out the other side of the hole. This phenomenon is known as **diffraction**.

Diffraction effects (Figure 3.16):
● are greater for waves of long wavelength
● are greater for small holes

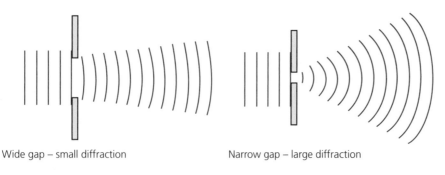

Wide gap – small diffraction

Narrow gap – large diffraction

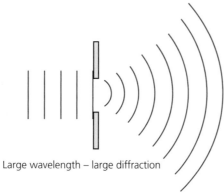

Large wavelength – large diffraction

Figure 3.16 Diffraction effects

Appearance of the diffraction pattern from a single slit

REVISED

When light passes through a single slit diffraction occurs. The variation of intensity with angle of diffraction (θ) is shown in Figure 3.17.

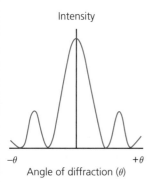

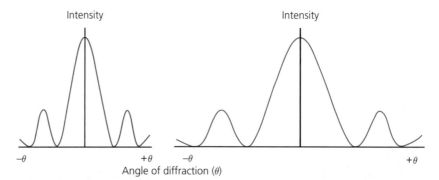

Figure 3.17 Single-slit diffraction

Figure 3.18 Diffraction and wavelength

Figure 3.18 shows the effect of a change of wavelength on the diffraction pattern.

Blue light — short wavelength, giving a narrow diffraction pattern.

Red light — long wavelength, giving a broad diffraction pattern.

If the slit is narrowed the diffraction pattern will become broader for a given wavelength, and it will narrow if the slit is made wider.

If white light is used, a spectrum is formed at each maximum in the diffraction pattern.

Now test yourself

TESTED

9 Why is the diffraction of light much more difficult to observe than the diffraction of microwaves?
10 Can diffraction occur with longitudinal waves as well as with transverse waves?

Answers on p. 216

The diffraction grating

REVISED

If a number of parallel, narrow slits are made the result is known as a diffraction grating (Figure 3.19). Those in use in schools have typically between 80 and 300 slits per mm. The distance between the centres of adjacent slits is called the **grating spacing** (d). If there are 100 slits per mm the width of one slit is 0.01 mm, or 10^{-5} m.

The diffracted images produced by diffraction gratings are both sharper and more intense than those produced by a single slit. The intensity in any given direction is the sum of those due to each slit (Figure 3.19).

The slits of a diffraction grating are usually called lines.

Formula for a diffraction grating

When a parallel beam of light falls on a diffraction grating a number of diffraction maxima are formed. These are formed at angles of θ to the normal of the grating. The formula for these maxima is:

diffraction grating maximum, $n\lambda = d \sin \theta$

where $n = 0, 1, 2, 3\ldots$

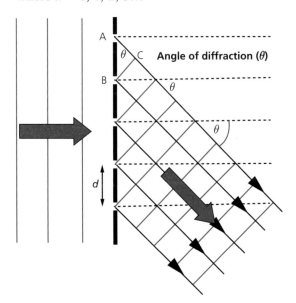

Figure 3.19 Diffraction grating

The number n is known as the **order of the spectrum** so that, for example, a first-order spectrum is formed for $n = 1$, and so on. If light of a single wavelength, such as that from a laser, is used, then a series of sharp maxima occur — one maximum to each order of the spectrum. A white light source gives a series of spectra with the light of the shortest wavelength having the smallest angle of diffraction.

The number of orders of spectra visible with a given grating depends on the grating spacing, with more spectra being visible with coarser gratings. This is because the maximum angle of diffraction is 90° and therefore the maximum value for $\sin\theta = 1$.

> **Exam tip**
>
> The number of lines per metre on a diffraction grating is the inverse of the grating spacing.

> **Example**
>
> Calculate the wavelength of the monochromatic light where the second-order image is diffracted through an angle of 25° using a diffraction grating with 300 lines per millimetre.
>
> **Answer**
>
> grating spacing, $d = \dfrac{10^{-3}}{300\,\text{m}} = 3.3 \times 10^{-6}\,\text{m}$
>
> wavelength, $\lambda = \dfrac{d\sin 25}{2} = \dfrac{3.3 \times 10^{-6} \times 0.42}{2} = 6.97 \times 10^{-7}\,\text{m} = 697\,\text{nm}$

> **Typical mistake**
>
> Forgetting that the diffraction occurs on both sides of the axis of the system.

Now test yourself

TESTED

11 A diffraction grating has 250 lines per mm. Calculate:
   (a) the angle of diffraction for the first order image for light of wavelength 550 nm
   (b) the highest order possible with this grating at this wavelength
   (c) the number of images of the source
12 How could you tell the difference between a CD and a DVD simply by looking at the diffraction pattern produced by a white light source? Explain your answer.

Answers on p. 216

Applications of diffraction gratings

Diffraction gratings are very useful tools for the study of spectra. They are often used to determine the composition of an incandescent source and are particularly useful in the analysis of the material of stars. Diffraction gratings are cheaper and easier to use than glass prisms and the spectra produced can be made large by increasing the number of lines per metre.

Using a diffraction grating to observe a spectrum
The grating is illuminated with light from a discharge lamp incident at right angles. Light is diffracted from all the slits and the resulting waves interfere with each other to give maxima and minima. The angle of diffraction for each maximum and minimum depends on the wavelength of the light.

Refraction at a plane surface

When light passes from one medium to another of different refractive index its speed changes (Figure 3.20). This change of speed depends on the refractive index of the two materials. It moves more slowly in a material of higher refractive index than it does in a material of low refractive index. This means that it will refract at a boundary between two media of different refractive index.

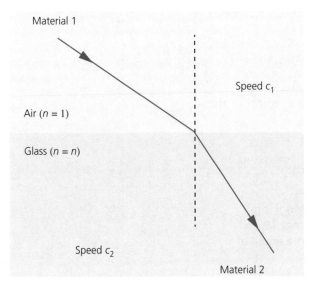

Material 1

Air ($n = 1$)

Glass ($n = n$)

Speed c_1

Speed c_2

Material 2

Figure 3.20 Refraction and change of speed

$$\text{refractive index, } n \text{ of material 2} = \frac{c_1}{c_2}$$

Example

A beam of light passes from air (refractive index 1.00) into diamond (refractive index 2.42), the speed of light in air being $3.0 \times 10^8\,\text{ms}^{-1}$. Calculate the speed of light in diamond.

Answer

$$\text{speed of light in diamond} = \frac{3 \times 10^8}{2.42} = 1.24 \times 10^8\,\text{ms}^{-1}$$

Refraction at a boundary between two different substances

The change of speed of a beam of light passing from material 1 to material 2 causes the light to refract. The amount of refraction obeys Snell's law of refraction (Figure 3.21).

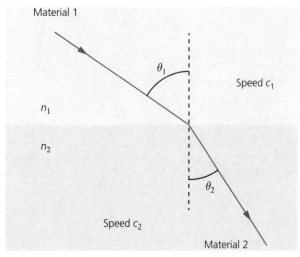

Figure 3.21 Refraction at a boundary

If the absolute refractive indices of the materials are n_1 and n_2 respectively, according to Snell's law:

$$n_1 \sin \theta_1 = n_2 \sin \theta_2$$

Example

A beam of light passes from water (refractive index 1.33 (n_1)) into diamond (refractive index 2.42 (n_2)). If the angle of incidence (θ_1) in water at the water–diamond boundary is 35°, calculate the angle of refraction (θ_2) in the diamond.

Answer

Using:

$$n_1 \sin \theta_1 = n_2 \sin \theta_2$$

$$\sin \theta_2 = \sin \theta_1 \frac{n_1}{n_2} = \sin 35 \frac{1.33}{2.42} = 0.574 \times 0.55 = 0.32$$

Therefore:

angle of refraction (θ_2) = 18.4°

Exam tip

The absolute refractive index of a material is its refractive index compared with a vacuum and is the value usually quoted simply as the refractive index.

Exam practice answers and quick quizzes at **www.hoddereducation.co.uk/myrevisionnotes**

Now test yourself

13 A beam of light travelling in glycerol passes into a diamond (refractive index 2.42). Calculate the refractive index of glycerol. The angle of incidence in glycerol is 35° and the angle of refraction in diamond is 20.4°.

Answer on p. 216

Total internal reflection

When light passes from a material such as water into one of lower refractive index such as air there is a maximum angle of incidence in the water that will give a refracted beam in the air, that is, the angle of refraction is 90°. The angle of incidence in the material of higher refractive index corresponding to an angle of refraction of 90° in the material of lower refractive index is known as the **critical angle**, c (Figure 3.22(a)).

If this angle of incidence is exceeded, *all* the light is reflected back into the material of higher refractive index. This is called **total internal reflection** (Figure 3.22(b)) and the normal laws of reflection are obeyed.

$$\sin \theta_c = \frac{n_2}{n_1}$$

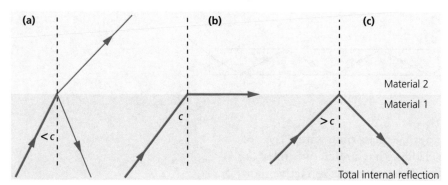

Figure 3.22 **The critical angle and total internal reflection**

> **Exam tip**
>
> Total internal reflection only occurs when the light travelling in one material meets a boundary with a material that has a lower refractive index.

Example

The refractive indices from air to glass and from air to water are 1.50 and 1.33 respectively. Calculate the critical angle for a water–glass surface.

Answer

The refractive index for light passing from water to glass [$_w n_g$] is given by:

$$_w n_g = \frac{n_g}{n_w} = \frac{1.5}{1.33} = 1.13$$

Therefore the critical angle (c) can be found from:

$$_w n_g = \frac{1}{\sin c}$$

And so:

$$\sin c = \frac{1}{1.13} = 0.89$$

Therefore, $c = 62.9°$

Now test yourself

14 Calculate the critical angle for a water–diamond interface if the refractive indices are 1.33 and 2.42 respectively.

15 Show that it is impossible for a beam of light to enter one face of a cubical glass block and leave by the adjacent face. (refractive index of the glass = 1.5)

Answers on p. 216

Fibre optics

REVISED

An important application of total internal reflection is in fibre optics. Light is shone along a thin glass fibre and if it hits the glass–air boundary at more than the critical angle it reflects along inside the fibre. These fibres are normally around 125 μm (0.125 mm) in diameter (similar to that of a human hair). The core diameter (see later) is around 50 μm.

The transmission of light down a glass fibre is of enormous importance in communications. Glass fibres are cheap, light in weight compared with copper wire and light can be modulated to carry an enormous amount of information. Figure 3.23 shows the situation for a single glass fibre in air.

Figure 3.23 Single glass fibre

Figure 3.24 shows the situation for a single glass fibre with a layer of glass cladding surrounding it. The cladding has a refractive index (n_1) significantly greater than air but slightly less than the refractive index of the core of the fibre (n_2).

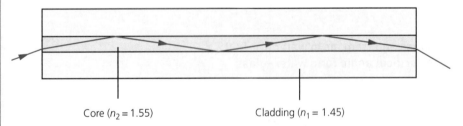

Core ($n_2 = 1.55$) Cladding ($n_1 = 1.45$)

Figure 3.24 Single glass fibre with cladding

The refractive indices from air to the glass of the core and from air to the glass of the cladding are 1.55 and 1.45 respectively. Calculate the critical angle for a water–glass surface.

Answer

The refractive index for light passing from the core to the cladding is:

$$\frac{n_{CORE}}{n_{CLADDING}} = \frac{1.55}{1.45} = 1.069$$

Therefore, the critical angle, c, can be found from:

$$_{CORE}n_{CLADDING} = \frac{1}{\sin c}$$

So:

$$\sin c = \frac{1}{1.069} = 0.935$$

Therefore, $c = 69.3°$

Material dispersion

If a pulse of white light is sent down a fibre with a single layer of cladding **material dispersion** occurs, with different wavelengths taking different times to travel down the fibre. This is because of the different refractive indices of the core for different wavelengths of light. As a result, the pulse spreads and the effect is known as **pulse broadening**.

Modal dispersion

Another problem is the different time of transmission between rays that make different angles with the axis of the fibre. This is known as **modal dispersion** (Figure 3.25) and can be reduced by cladding the fibre. The greater critical angle when cladding is used means that only rays of light that have a large angle of incidence with the 'walls' of the fibre, and so make a small angle with the axis of the fibre, will be transmitted along it. Other light will refract out of the fibre. This reduces the difference in time of transmission and so the spread of information with time is also reduced.

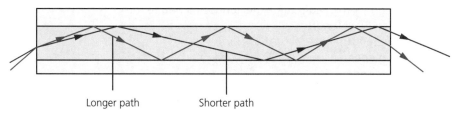

Longer path   Shorter path

Figure 3.25 The effect of modal dispersion

The distance that information can be sent down the fibre will be limited both by absorption in the glass of the fibre and by the effects of dispersion if light that is not truly monochromatic is used.

Now test yourself   TESTED ☐

16 Explain briefly why the use of non-monochromatic light is a problem for the transmission of data along an optical fibre.

Answer on p. 216

Exam practice

Take the velocity of light in free space to be $3 \times 10^8 \, \mathrm{m\,s^{-1}}$ where needed.

1 A cello D string has a length of 70 cm and a first harmonic of frequency 73.3 Hz.
   (a) What is the frequency of the second harmonic? [3]
   (b) Where would the cellist have to touch the string to produce this harmonic? [1]
   (c) Calculate the tension in the cello string. [3]
   (mass per unit length = $1.5 \times 10^{-3} \, \mathrm{kg\,m^{-1}}$)

2 A stretched wire with its ends firmly clamped has a first harmonic of frequency 1000 Hz. What will be the frequency of the first harmonic if the tension of the wire is increased by 2%?
   A 980 Hz
   B 1040 Hz
   C 1020 Hz
   D 1000 Hz [1]

3 (a) Why can't you get a static interference pattern with two light bulbs, while it is possible with two loudspeakers? [2]
   (b) A Young's double-slit experiment is carried out using green light. Describe and explain what will happen to the interference fringes produced if:
      (i)   red light is used instead [1]
      (ii) blue light is used instead [1]
      (iii) the two slits are moved closer together [1]
      (iv) the two slits are moved further apart [1]
      (v)   white light is used [1]
      (vi) one of the slits is covered up [2]
      (vii) the slits are made narrower. [1]

4 Light from a fluorescent lamp is found to consist of only two wavelengths, 450 nm and light of a longer wavelength. When the light is passed through a given diffraction grating it is found that the third order for the shorter wavelength has the same diffracted angle as the second order for the longer wavelength. Calculate:
   (a) the wavelength of the light with the longer wavelength [3]
   (b) the grating spacing if the angle of diffraction in this case is 20°. [2]

5 Water waves of wavelength 4 m meet a narrow entrance to a harbour. If the entrance is 10 m wide calculate the separation between the central maximum and the first minimum on the beach if the distance from the harbour mouth to the beach is 250 m. [2]

6 A beam of light of wavelength 600 nm hits a glass–air interface at an angle of 40°. If the velocity of light in glass is $2 \times 10^8 \, \mathrm{m\,s^{-1}}$ find:
   (a) the angle of refraction [2]
   (b) the wavelength of the light in the glass. [2]
   (c) A thin layer of glass of refractive index 1.45 is now added to the original glass surface. Calculate the critical angle for the interface between the two pieces of glass. [3]

7 When visible light passes from air into glass the radiation experiences a change in:
   A frequency but not in speed and not in wavelength
   B frequency and speed but not in wavelength
   C wavelength and frequency but not in speed
   D wavelength and speed but not in frequency. [1]

8 Explain carefully the effect on the spectrum observed by a plane transmission diffraction grating if the ruled face is presented to the incident light rather than the un-ruled face. [2]

Answers and quick quiz 3 online

ONLINE

Summary

You should now have an understanding of:

- progressive waves and stationary waves — waves are formed by oscillations of particles or fields
- how longitudinal waves show oscillations along the direction of wave propagation while transverse waves show oscillations at right angles to the direction of wave propagation
- the principle of wave superposition — the formation of points of no vibration (nodes) and points of maximum vibration (antinodes) on a stretched string
- formation of stationary waves by two travelling waves moving 'through' each other in opposite directions and combining
- interference — the overlapping a two systems of waves giving a pattern showing maxima and minima
- double-slit interference — this gives fringes of one colour using monochromatic light and a series of spectra using white light
- diffraction — this is the spreading of waves through a hole or round an obstacle; it is greater for small obstacles and holes and for long-wavelength waves.
- the use of diffraction grating to give finer and brighter spectra
- refraction of light at a plane surface, producing a change in direction of the refracted light; the speed in a material of high refractive index is less than that in a material of low refractive index
- total internal reflection — this only occurs when light moving in one medium meets a boundary with one with a lower refractive index
- fibre optics — the transmission of light and/or microwaves along a glass fibre; cladding is used to reduce both material and modal dispersion

Scalars and vectors

The quantities measured in physics can be divided into two groups, **scalars** and **vectors**.

> **Scalars** are quantities that have magnitude (size) only. Examples of scalars are length, speed, mass, density, energy, power, temperature, charge and potential difference.
>
> **Vectors** are quantities that have direction as well as magnitude. Examples of vectors are displacement, force, torque, velocity, acceleration, momentum and electric current.

Scalars can be added together by simple arithmetic but when two or more vectors are added together their direction must be taken into account as well.

A vector can be represented by a line, the length of the line being the magnitude of the vector and the direction of the line the direction of the vector.

Addition of vectors

REVISED

When two or more vectors are added the resulting sum of the vectors is called the **resultant vector** or simply the **resultant**.

Vectors acting in the same line

Two or more vectors acting in the same direction can be added as if they were scalars. For example the sum, or resultant, of the three forces shown in Figure 4.1(a) is 100 N acting right to the left while in Figure 4.1(b) it is 700 N left to right.

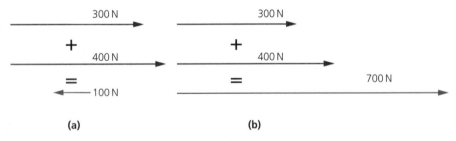

(a)   (b)

Figure 4.1 Vectors acting in the same line

Vectors acting in different directions

If the two vectors acting at a point are not acting along the same line the resultant can be found by either using a scale diagram or by calculation. The vectors are drawn nose to tail and the resultant closes the triangle. (Note: calculations will be limited to two vectors at right angles (Figure 4.2(a)).

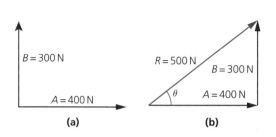
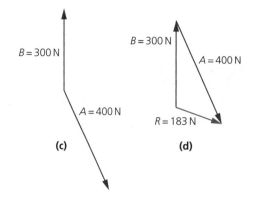

Figure 4.2 Vectors acting at an angle

Examples

Finding the resultant using a scale diagram

Using the same magnitude for the two vectors as those already considered we draw a scale diagram in both magnitude and direction. The resultant, R (= 500 N in this case) is the vector that closes the triangle (Figure 4.2b)

The direction of the resultant can be found by measuring angle θ. In this case $\theta = 37°$.

Finding the resultant by calculation

The resultant of the two vectors can also be found by calculation.

$$\text{resultant, } R = \sqrt{400^2 + 300^2} = \sqrt{250\,000} = 500\,\text{N}$$

The direction of R can be found from:

$$\tan\theta = \frac{300}{400} = 0.75$$

So, $\theta = 36.9°$

Exam tip

Notice that the original two vectors (shown blue in Figure 4.2) follow each other round the triangle (nose to tail) to give the **resultant**, the red vector (R), and that this resultant acts in the opposite direction round the triangle.

Exam tip

Always choose a sensible scale when drawing scale diagrams of vectors.

An example of two vectors not acting at right angles to each other is shown in Figure 4.2(c) and Figure 4.2(d).

Now test yourself

TESTED

1 An aircraft is flying on an initial bearing of 0° at 350 m s⁻¹ with a wind blowing west to east at 50 m s⁻¹. Find the true speed of the plane and the direction in which it travels. You should use both the scale diagram and calculation methods to find your answers.

Answer on p. 217

Components of vectors

REVISED

The effectiveness of the vector along a specified direction is called the **component** of the vector along this direction. Finding the components of a vector, usually along two perpendicular directions, is called the **resolution** of a vector.

The component of a vector along any direction is the magnitude of the vector multiplied by the cosine of the angle between the vector and the line.

The horizontal component of the vector **F** shown in Figure 4.3 (a) is $F\cos A$, while Figure 4.3(b) shows the components of a vector in two perpendicular directions. These are known as the rectangular components of the vector.

Component in the x direction: $\mathbf{F}_x = F \cos A$

Component in the y direction: $\mathbf{F}_y = F \cos(90 - A) = \mathbf{F} \sin A$

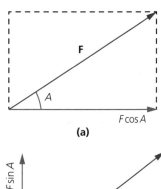

Typical mistake

Mixing up the angles when calculating components.

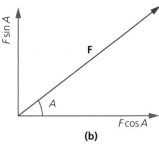

(a)

Example

A railway truck is pulled along the rails by a rope that makes an angle of 35° with the track. If the force (**F**) in the rope is 1500 N calculate the components of the force:

(a) perpendicular to the rails

(b) parallel to the rails.

Answer

(a) component of **F** perpendicular to the rails = $F \sin 35 = 1500 \times 0.574$
$= 860$ N

(b) component of **F** parallel to the rails = $F \cos 35 = 1500 \times 0.819 = 1229$ N

Figure 4.3 Components of a vector

Now test yourself

2 A tug pulls an ocean liner from its moorings. The cable from the bow of the liner to a tug makes an angle of 15° with the horizontal. If the force in the cable is 2000 N what are the horizontal and vertical components of this force?

Answer on p. 217

The inclined plane

The components of the forces acting on a point object (P) on an inclined plane are shown in the Figure 4.4. The actual forces are the weight of the object represented by the black vector and the reaction of the plane, shown by the blue vector. The components of the weight are shown in red.

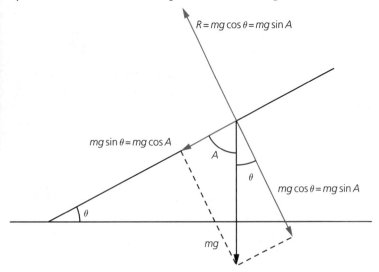

Figure 4.4 The inclined plane

Equilibrium due to two or more forces

A body acted on by two or more forces is in equilibrium when it has no tendency to move. This means that:

A body is in equilibrium when acted on by two or more coplanar forces if the resultant of these forces is zero and the two forces pass though one point.

The resultant can be shown to be zero by calculating the components of each force in two perpendicular directions. Alternatively a scale diagram can be drawn. The resultant is zero when the forces form a closed triangle.

This can mean that an object in equilibrium can be at rest or moving with a constant velocity.

> **Example**
>
> Three forces of 100 N, 150 N and 200 N act on a body, as shown in Figure 4.5. Show that the resultant of these three forces is zero and that therefore the body is in equilibrium.
>
>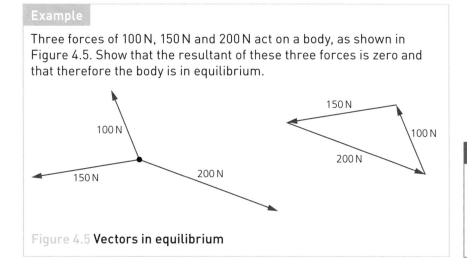
>
> Figure 4.5 **Vectors in equilibrium**

Exam tip

For the body to be in equilibrium — no translation or rotation — the forces acting on it must pass through one point.

Note that, in Figure 4.5, the vectors are drawn 'nose to tail'.

Moments

Moment of a force about a point

REVISED

If a force acts on an object, and the line of action of the force does not pass through the centre of mass of the object then the force will exert a turning effect on the object and it will rotate. The larger the force and the further the line of action from the centre of mass the greater the turning effect of the force will be.

This turning effect is called the **moment of the force**.

> **moment of a force** about a point = magnitude of force (F) × perpendicular distance (d) from the point to the line of action of the force:
>
> $$\text{moment} = Fd$$

Moments are measured in newton metres (N m).

Example

A light rod 80 cm long is pivoted about one end and supported by a vertical thread at the far end so that the rod makes an angle of 55° with the vertical. Calculate the moment of the force about the pivot if the force in the thread is 25 N.

Answer

See Figure 4.6.

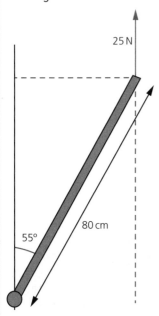

25 N

80 cm

55°

Figure 4.6 **Force causing a moment**

moment = force × perpendicular distance from the pivot to the line of action of the force

= 25 × 0.80 sin 55 = 16.4 N m

A body is in equilibrium when acted on by two or more coplanar forces if the resultant of these forces and their moments is zero.

Now test yourself

TESTED

3 A door has a handle that is 0.8 m from the hinge. What is the moment of the following forces applied to the handle about the hinge?

(a) 12 N at right angles to the door

(b) 20 N at 70° to the door

(c) 100 N at 25° to the door

Answers on p. 217

Couples and torque

If two equal and opposite forces, whose lines of action are not the same, act on a body, then they only produce a rotation of the body but no translation. This effect is called a **couple**.

A **couple** is composed of two forces that:
- are equal
- are anti-parallel (parallel but in opposite directions)
- do not pass through the same point.

Since a torque is caused by two forces rather than one the magnitude of the turning effect of a couple is called the **torque**.

The **torque** of a couple is the product of one of the forces and the perpendicular distance between the lines of action of the forces.

$$\text{torque} = \frac{F \times d}{2} \times 2 = Fd$$

Example

Two forces, each of 30 N, act on a rod pivoted at its centre, as shown in Figure 4.7. The ends of the forces on the rod are 60 cm apart. Calculate the torque produced.

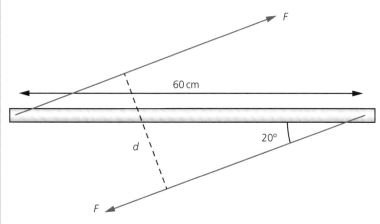

Figure 4.7 **A torque**

Answer

torque = force × perpendicular distance between the lines of action of the forces

torque = 30 × 0.60 sin 20 = 6.2 N m

Now test yourself

4 One of the powered wheels of a car travelling at constant velocity has a torque of 140 N m applied to it by the axle that drives the car. If the wheel is 0.45 m in diameter, calculate the driving force provided by this wheel.

Answer on p. 217

The principle of moments and equilibrium

When an object is balanced (in equilibrium) the sum of the clockwise moments is equal and opposite to the sum of the anticlockwise moments (Figure 4.8).

An example of the principle of moments is shown in Figure 4.8.

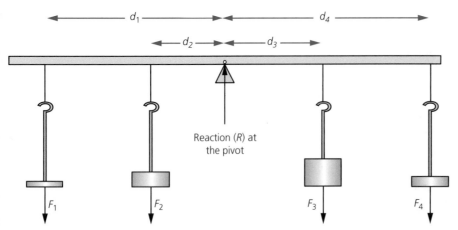

Figure 4.8 Principle of moments

$$F_1d_1 + F_2d_2 = F_3d_3 + F_4d_4$$

When an object is in equilibrium the sum of the vertical forces is zero.

$$F_1 + F_2 + F_3 + F_4 - R = 0$$

> **Exam tip**
>
> Notice the minus sign – R acts in the opposite direction from F_1, F_2, F_3 and F_4.

> **Example**
>
> In Figure 4.9, let $L = 100\,cm$, $F_1 = 20\,N$, $F_3 = 10\,N$, $d_1 = 10\,cm$, $d_2 = 70\,cm$, $d_3 = 45\,cm$

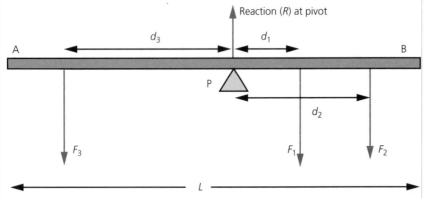

> **Exam tip**
>
> The moments can be taken about any point on an object. However, it may be more convenient to use one point than another if an unknown force passes through that point.

Figure 4.9 **Example of the principle of moments**

Find the value of F_2 such that the beam is in equilibrium.

Answer

Take moments about the pivot:

clockwise moments $= (20 \times 0.1) + (F_2 \times 0.7) = 10 \times 0.45 =$ anticlockwise moments

$2 + (F_2 \times 0.7) = 10 \times 0.45$

$$F_2 = \frac{4.5 - 2}{0.7} = 3.6\,N$$

But:

$F_1 + F_2 + F_3 - R = 0$

and so:

$R = 20 + 3.6 + 10 = 33.6\,\text{N}$

Now test yourself

5 A 3 m long uniform plank of mass 6 kg is fixed to a wall by a pivot at P. It is supported by wire which makes an angle of 55° with the plank, and is fixed to it 0.6 m from the end furthest from the wall, as shown in Figure 4.10. $(g = 9.8\,\text{m s}^{-2})$

Using the principle of moments calculate the tension (T) in the wire.

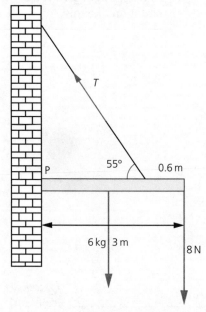

Figure 4.10 **Tension in a wire**

6 A lightweight beam 5.5 m long is fixed to a pivot. A load of 600 N is hung from one end and the beam is held in equilibrium by a vertical force F, as shown in Figure 4.11.

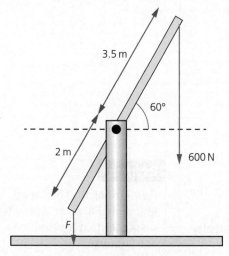

Figure 4.11 **A lightweight beam**

(a) Using the principle of moments calculate the value of F.
(b) How does F change if the angle of the beam is changed?

Answers on p. 217

Centre of gravity and centre of mass

The weight of an object may be taken as acting at one point known as the **centre of gravity**. You could think of that point as the position where all the mass of the object is concentrated.

> **The resultant moment about the centre of gravity of any object must be zero.**

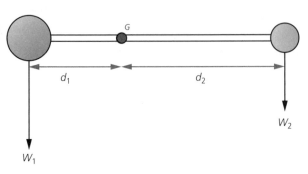

Figure 4.12 **Centre of mass of an unequal dumbbell**

The dumbbell arrangement in Figure 4.12 shows this clearly ($W_1 = m_1 g$; $W_2 = m_2 g$).

> **anticlockwise moment = $W_1 \times d_1$; clockwise moment = $W_2 \times d_2$**

If all the mass was replaced by a mass of $M (= m_1 + m_2)$ at G there would be no turning effect about G and therefore the resultant moment must be zero.

> **Exam tip**
>
> If the gravitational field is uniform over the size of the object then the centre of mass and the centre of gravity of the object will coincide. However if this is not true then they will be in different places. Close to a black hole, where the gravitational field changes rapidly, might give you a situation like this.

Motion along a straight line

Displacement, speed, velocity and acceleration

> **Speed** is defined as the rate of change of distance with time, while **velocity** is defined as the rate of change of displacement with time.

If an object is displaced by a small amount Δs in a small time Δt then its velocity v is given by the equation:

$$\text{velocity, } v = \frac{\Delta s}{\Delta t}$$

> **Displacement** is distance measured in a particular direction and **velocity** is speed measured in a particular direction. The units normally used for both speed and velocity are m s⁻¹.
>
> Acceleration occurs when an object changes its velocity with time. The units normally used for acceleration are m s⁻²:
>
> $$\text{acceleration, } a = \text{rate of change of velocity} = \frac{\Delta v}{\Delta t} = \frac{\text{change in velocity}}{\text{time}}$$

> **Exam tip**
>
> Remember that distance and speed are scalars, while displacement and velocity are vectors.

Using graphical methods to understand motion

Displacement–time graphs

Figure 4.13 shows examples of displacement–time graphs. Uniform velocity is shown by straight lines and acceleration is shown by a curved line. The gradient at a point on a displacement–time curve is the instantaneous velocity at that point.

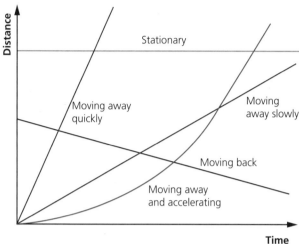

Figure 4.13 Displacement–time graphs

Velocity–time graphs

Figure 4.14 shows examples of velocity–time graphs. Uniform acceleration is shown by straight lines. Varying acceleration is shown by a curved line. The gradient at a point on a velocity–time curve is the instantaneous acceleration at that point.

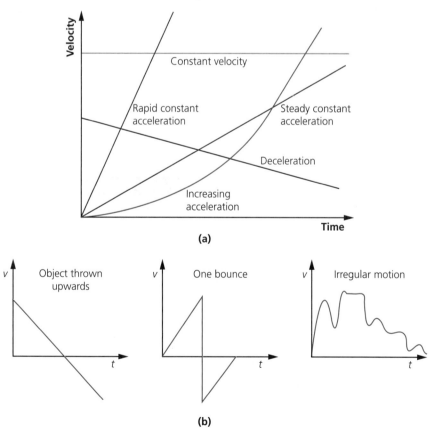

Figure 4.14 Velocity–time graphs

The area beneath a velocity–time graph is a measure of the displacement of the object (Figure 4.15).

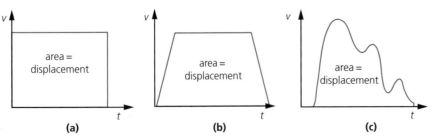

Figure 4.15 Area under velocity–time graphs

Example

Figure 4.16 shows the motion of a car.

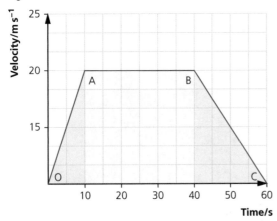

Figure 4.16 **Velocity–time graph example**

- O to A — the velocity increases steadily from $0\,\mathrm{m\,s^{-1}}$ to $20\,\mathrm{m\,s^{-1}}$ in 10 seconds.
- A to B — the velocity stays the same at $20\,\mathrm{m\,s^{-1}}$ for the next 30 s.
- B to C — the velocity decreases to $0\,\mathrm{m\,s^{-1}}$ in 20 s.

Using:

distance = average velocity × time

Distance travelled OA = 10 × 10 = 100 m

Distance travelled AB = 20 × 30 = 600 m

Distance travelled BC = 10 × 20 = 200 m

Total distance travelled OC = 900 m

This is represented by the area under the line OABC.

Velocity–time graphs showing non-uniform acceleration

Problems of non-uniform acceleration can be solved by using graphical methods. One example is the 100 m sprint shown in Figure 4.17.

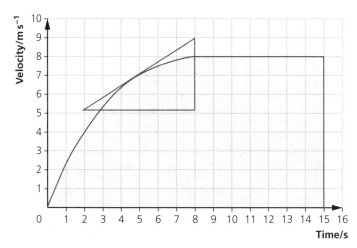

Figure 4.17 Non-uniform acceleration

The graph in Figure 4.17 shows the motion of a schoolgirl sprinter running the 100 m. It shows that her acceleration was not uniform but varied over the first 9 s of the race, after which it was zero.

The shaded area represents the distance she travelled in 15 s — in this case 100 m.

Her acceleration during the early part of the race can be found from the gradient of the curve at that point. For example, 5 s after the start her acceleration ($\Delta v/\Delta t$) was approximately $3.8/6 = 0.63\,\text{m s}^{-2}$.

Now test yourself

7 The graph in Figure 4.18 shows a runner during part of a race. Using the graph find:

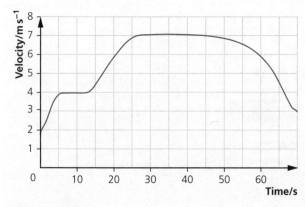

Figure 4.18

(a) the distance covered by the runner between 20 s and 60 s
(b) the velocity of the runner at $t = 15\,\text{s}$
(c) the acceleration of the runner at $t = 55\,\text{s}$.

Answers on p. 217

Equations for uniform acceleration

If the acceleration of an object is uniform, the following equations apply to its motion:

$$\text{average velocity} = \frac{v + u}{2} \text{ and } s = \frac{u + v}{2}t$$

$$\text{acceleration, } a = \frac{v - u}{t} \text{ or } v = u + at$$

$$s = ut + \tfrac{1}{2}at^2$$

$$v^2 = u^2 + 2as$$

> **Exam tip**
>
> Remember that these equations only apply to accelerated motion when the acceleration is uniform.

where u is the initial velocity, v the final velocity, a the acceleration, t the time taken and s the displacement.

Example 1

A dragster starts from rests and accelerates at $25\,\text{m s}^{-2}$ for $4\,\text{s}$. Calculate:
(a) the final velocity
(b) the distance travelled.

Answer

(a) $v = u + at = 0 + (25 \times 4) = 100\,\text{m s}^{-1}$
(b) $s = \tfrac{1}{2}at^2 = 0.5 \times 25 \times 16 = 200\,\text{m}$

Example 2

A ball travelling at $20\,\text{m s}^{-1}$ is hit by a bat and returned along its original path but in the opposite direction at $35\,\text{m s}^{-1}$. If the ball was in contact with the bat for $0.02\,\text{s}$ calculate:
(a) the acceleration of the ball during the hit
(b) the distance moved by the ball during the hit.

Answer

(a) $a = \dfrac{v - u}{t} = \dfrac{35 - (-20)}{0.02} = \dfrac{55}{0.02} = 2750\,\text{m s}^{-2}$
(b) $v^2 = u^2 + 2as$

$s = \dfrac{35^2 - 20^2}{2 \times 2750} = 0.15\,\text{m}$

Now test yourself

8 Starting from rest a car travels for 2 minutes with a uniform acceleration of $0.3\,\text{m s}^{-2}$ after which its speed is kept constant until the car is brought to rest with a uniform retardation of $0.6\,\text{m s}^{-2}$. If the total distance travelled is $4500\,\text{m}$ how long did the journey take?
9 An electron in a TV tube emitted from rest from a hot cathode reaches a velocity of $10^7\,\text{m s}^{-1}$ when it passes the anode. Find the acceleration of the electron if the cathode and anode are separated by $3\,\text{cm}$.

Answers on p. 217

Exam practice answers and quick quizzes at **www.hoddereducation.co.uk/myrevisionnotes**

Acceleration due to gravity

The vertical acceleration in the Earth's gravitational field is due to the gravitational attraction of the Earth, and is called the acceleration due to gravity (g) or the **acceleration of free fall**. The value of g close to the Earth's surface is about $9.81\,\text{ms}^{-2}$ (often simplified to $9.8\,\text{ms}^{-2}$ or even $10\,\text{ms}^{-2}$).

Since this acceleration is produced by the gravitational field of the Earth it may also be called the **gravitational field strength** (units Nkg^{-1}).

If an object is dropped and falls through a height h in t seconds its acceleration is:

$$\text{gravitational acceleration } g = \frac{2h}{t^2}$$

If air resistance is neglected the acceleration in free fall is the same for all objects. This was suggested by Galileo Galilei in the seventeenth century.

> **Exam tip**
>
> You should always use the values of constants such as g that are given in your exam question paper.

Example

A stone falling from rest falls half its total path in the last second before it strikes the ground. From what height was it dropped?

Answer

For the complete path:

$h = \frac{1}{2}gt^2$

For the top half of its path:

$\dfrac{h}{2} = \frac{1}{2}g(t-1)^2$

So:

$h = \frac{1}{2}gt^2 = g(t^2 - 2t + 1)$

Therefore:

$t^2 = 2t^2 - 4t + 2$

and so:

$t^2 - 4t + 2 = 0$

This can be solved to give $t = 3.41\,\text{s}$ or $0.59\,\text{s}$. This last one is impossible since it fell half the distance in the last second.

Therefore:

$h = \frac{1}{2}gt^2 = \frac{1}{2} \times 9.8 \times 3.41^2 = 57.1\,\text{m}$

Now test yourself

10 A stone is dropped from a cliff.
   (a) How far will it have fallen in 4 s?
   (b) What will its velocity be at that point?
   (c) What is the average velocity of the stone during the 4 s? (Use $g = 9.8\,\text{ms}^{-2}$ and ignore air resistance.)

Answer on p. 217

Required practical 3

Determination of g

The value of the gravitational acceleration (gravitational field strength) at the Earth's surface may be found using a freefall method. A ball bearing is dropped from rest and the time (t) for it fall through a known height (h) is found. The measurements are repeated both for the original height and over a series of heights from 20 cm to 2 m.

The height fallen is measured with a ruler and the time of fall by a light gate or a mechanical gate mechanism.

$$\text{gravitational acceleration} = \frac{2h}{t^2}$$

Projectile motion

Objects projected vertically

REVISED

If an object is projected upwards with an initial vertical velocity of u, such that its time of flight is $2t$ (in other words the time to return to the ground again) the time to reach the top of its trajectory is t. The velocity (v) at the top of the trajectory is zero.

maximum height, $h = \frac{1}{2}gt^2 = \frac{u^2}{2g}$

velocity, $v = u + gt$

> **Exam tip**
>
> Remember that the acceleration of the projectile is $9.8\,\mathrm{m\,s^{-2}}$ towards the ground throughout the trajectory, even at the very top.

Example

A ball is thrown vertically upwards with an initial velocity of $30\,\mathrm{m\,s^{-1}}$.
Calculate:
(a) the maximum height reached
(b) the time taken for it to return to the ground.

($g = 9.8\,\mathrm{m\,s^{-2}}$)

Answer

(a) Using $v^2 = u^2 + 2as$:

$0 = 900 - 2 \times 9.8 \times s$

$19.6s = 900$

$s = 45.9\,\mathrm{m}$

(Notice that at the maximum height the vertical velocity is zero and that the acceleration due to gravity is negative since it acts to retard the ball.)

(b) Using $v = u + at$:

$30 = -30 + 9.8t$

$t = 6.1\,\mathrm{s}$

(Remember that the ball must return to the ground with the same speed with which it left it.)

Objects projected horizontally

Figure 4.19 shows the motion of an object projected horizontally in a gravitational field.

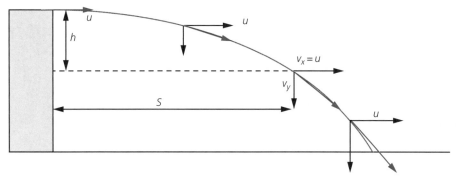

Figure 4.19 Projectile: horizontal projection

The object has:
- a motion in the horizontal direction — this is uniform velocity, since no forces act in this direction
- a motion in the vertical direction — this is uniformly accelerated motion due to the gravitational pull of the Earth, the vertical acceleration being the strength of the Earth's field ($g = 9.8\,\text{m s}^{-2}$). Remember that this always acts vertically downwards.

> **Exam tip**
>
> The motion of a projectile should be thought of in two separate parts — one horizontal and the other vertical.

The horizontal distance travelled (s) = horizontal velocity $\times$ time = $v_x t = ut$

The vertical distance travelled (h) = $u_y t + \tfrac{1}{2}(gt^2) = \tfrac{1}{2}(gt^2)$ since $u_y = 0$.

Velocity after a time t $v = (v_x^2 + v_y^2)^{1/2}$

Direction of motion after time t $\tan\theta = v_y/v_x$

Example

A ball is thrown horizontally with an initial velocity of $6\,\text{m s}^{-1}$ from an open window that is $4\,\text{m}$ above the ground. Calculate:
(a) the time it takes to hit the ground
(b) the distance from the wall where it hits the ground
(c) the velocity (magnitude and direction) 0.5 seconds after it is thrown.

(Ignore air resistance in your calculations and take $g = 9.8\,\text{m s}^{-2}$.)

Answer

(a) Using $h = \tfrac{1}{2}gt^2$:
$4 = \tfrac{1}{2} \times 9.8 \times t^2$
$t = 0.904\,\text{s} = 0.90\,\text{s}$
(b) $s = vt = 6 \times 0.904 = 5.42\,\text{m}$
(c) vertical velocity after $0.5\,\text{s} = 0 + gt = 9.8 \times 0.5 = 4.9\,\text{m s}^{-1}$
velocity after $0.5\,\text{s} = \sqrt{v_x^2 + v_y^2} = \sqrt{6^2 + 4.9^2} = 7.75\,\text{m s}^{-1}$
direction of motion: $\tan\theta = \dfrac{4.92}{6} = 0.82$ and so $\theta = 39.4°$

11 A crate is released from an aircraft that is flying horizontally, 1500 m above the ground, at a steady speed of 200 m s⁻¹.
   (a) What is its horizontal velocity:
      (i) 2 s after it was released?
      (ii) 5 s after it was released?
   (b) What is its vertical velocity:
      (i) 2 s after it was released?
      (ii) 5 s after it was released?
   (c) What is its velocity (magnitude and direction) 5 s after it was released?
   (d) How long will it take to reach the ground?
   (e) How far horizontally from the place where it was released will it hit the ground?

(g = 9.8 m s⁻²; air resistance can be ignored)

Answer on p. 217

Objects projected at an angle

REVISED

Consider an object projected with velocity u at an angle A to the horizontal (Figure 4.20).

vertical component of velocity = $a \sin A$

horizontal component of velocity = $u \cos A$

Vertical motion: $h = ut \sin A - \frac{1}{2}gt^2$

Horizontal motion: $s = ut \cos A$

The maximum range for a given velocity of projection is when $\sin 2A = 1$, that is, when $2A = 90°$ or when $A = 45°$.

$$\text{range} = \frac{u^2 \sin 2A}{g}$$

maximum height reached, $H = \frac{u^2 \sin^2 A}{2g}$

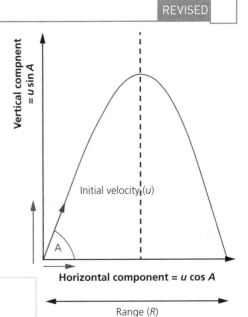

Figure 4.20 Object projected at an angle

Example

A stone is projected at an angle of 60° to the horizontal with a velocity of 30 m s⁻¹. (g = 9.8 m s⁻²)

Calculate:
(a) the highest point reached
(b) the range
(c) the time taken for the flight
(d) the height of the stone at the instant that the path makes an angle of 30° with the horizontal.

Answer

(a) highest point = $\dfrac{30^2 \sin^2 60°}{2g} = \dfrac{900 \times 0.75}{19.6} = 344$ m

(b) range = $\dfrac{30^2 \sin 120°}{9.8} = \dfrac{900 \times 0.866}{9.8} = 79.6$ m

(c) time of flight = $\dfrac{2 \times 30 \sin 60°}{9.8} = 5.3$ s

(d) At the point when the path makes an angle of 30° to the horizontal:

$\tan 30° = \dfrac{\text{vertical component of velocity}}{\text{horizontal component of velocity}} = \dfrac{\text{vertical component}}{30 \cos 60}$

The vertical component (v) is given by the formula:
$v^2 = 30^2\sin^2 60° - (2 \times 9.8 \times h) = (900 \times 0.75) - 19.6h$
where h is the height reached at that point. Therefore:
$\tan 30° = \dfrac{675 - 19.6h}{15}$
$15 \times 0.58 = 675 - 19.6h$
$8.66 = 675 - 19.6h$
$19.6h = 666.3$
$h = 34.0\,\text{m}$

Now test yourself

TESTED

12 A projectile is fired from a siege engine at $35\,\text{m s}^{-1}$ at an angle of $60°$ to the horizontal:
   (a) What is its vertical velocity at the top of its flight?
   (b) What is its horizontal velocity at the top of its flight?
   (c) What is its acceleration at the top of its path?
   (d) What is its vertical velocity 5 s after it is fired?
   (e) What is its overall velocity 5 s after it is fired (magnitude and direction needed here).

Answer on p. 217

Friction

REVISED

To move one body over another that is at rest requires a force. This is needed both to change the momentum of the first body and also to overcome the frictional force between the two surfaces. The force needed to overcome the frictional force when the bodies are at rest is called the **limiting friction** (often called static friction).

The static frictional force between two surfaces depends on:
- **the nature of the two surfaces**
- **the normal reaction between them**

When the object is moving the friction between the two surfaces is usually less than the static friction. This is almost independent of the relative velocities of the two surfaces.

Lift and drag forces

REVISED

When a fluid is in motion the pressure within the fluid varies with the velocity of the fluid if the flow is streamlined.

The pressure within a fast-moving fluid is lower than that in a similar fluid at rest or moving slowly.

The shape of the cross-section of an aircraft wing is designed so that the velocity of the air above the wing is greater than that below it. A region of low pressure is created above the wing and so the aircraft experiences an upward force known as lift (Figure 4.21).

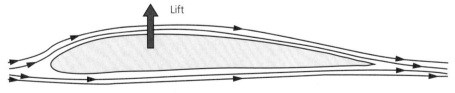
Lift

Figure 4.21 Airflow over an aircraft wing

Terminal speed

When any object falls through a fluid such as air it will experience a viscous drag.

As the object falls faster and faster the drag force increases. Eventually the drag force increases to a value where it is equal to the weight of the object and the body continues to fall at a steady speed. We call this the **terminal speed** of the object (Figure 4.22).

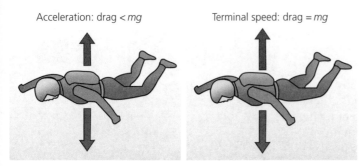

Acceleration: drag < *mg*   Terminal speed: drag = *mg*

Figure 4.22 Terminal speed

At the terminal speed:

viscous drag (air resistance) = weight of the object = *mg*

Figure 4.23 shows how the velocity of an object will increase with time as it falls through a viscous fluid. The acceleration starts with a value of *g* but falls to zero when the terminal speed is reached.

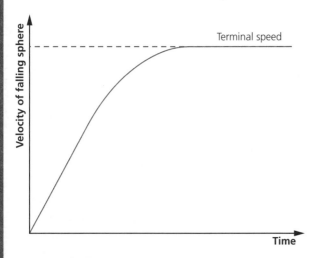

Figure 4.23 Terminal speed graph

Newton's laws of motion

Newton's first law of motion

Newton's first law of motion states that:

> **A body remains at rest or in a state of uniform motion unless acted on by a resultant force.**

Exam practice answers and quick quizzes at **www.hoddereducation.co.uk/myrevisionnotes**

At rest

Someone sitting on a stool may be at rest but they are acted on by two forces — their weight and the reaction of the stool. It is because these two forces are balanced and there is no resultant force that they stay still — i.e. at rest.

Uniform motion

This means no change of velocity; since velocity is a vector, this means at a steady speed in a straight line. As a skydiver falls out of a plane their speed increases — their weight is bigger than the drag — so there is a net force and Newton's first law does not apply. However, as the drag increases the two forces on them become equal and the skydiver falls with a constant velocity — a state of uniform motion.

Newton's second law of motion

REVISED

You need a force to change the motion of a body. The bigger the net force the greater the acceleration. (Remember that force is a vector and the direction of the forces acting on a body need to be considered.)

Newton's second law of motion suggests that:
- **acceleration is directly proportional to the accelerating force**
- **acceleration is inversely proportional to the mass of the body being accelerated**

$$\text{force} = \text{mass} \times \text{acceleration}$$

$$F = ma$$

The units for force are newtons (N), those for mass are kg and those for acceleration $\mathrm{m\,s^{-2}}$.

This law also gives us a good definition of the newton as a unit of force.

One newton is the force that will give a mass of 1 kilogram an acceleration of $1\,\mathrm{m\,s^{-2}}$.

Example

A 250 kg box is pulled along a smooth road (no frictional drag) using a force of 2000 N. What is the acceleration of the block?

Answer

Using $F = ma$:

$$a = \frac{F}{m} = \frac{2000}{250} = 8\,\mathrm{m\,s^{-2}}$$

Exam tip

Note that it is mass that is used in the equation, mass is measured in kg. Do not convert mass to a weight.

Now test yourself

TESTED

13 At the start of a 100 m race the rear foot of a sprinter can exert a force of some 1150 N on the starting blocks and the front foot an additional 800 N. If the sprinter is a man of mass 83 kg what is his initial acceleration?

14 If a rocket has a mass of 50 000 kg and its motors exert a thrust of 550 000 N what is the initial vertical acceleration off the rocket? ($g = 9.8\,\mathrm{m\,s^{-2}}$)

Answers on p. 217

Newton's third law of motion

This states that:

> **If a force acts on one body, an equal and opposite force acts on another body.**

or

> **Action and reaction are equal and opposite.**

This law can be checked by fixing two spring-loaded trucks together on a linear air track. When the spring is released they *both move off*, showing that there is a force on *both*, with the acceleration of each truck depending on its mass.

The two forces mentioned in Newton's third law are known as a 'Newton pair' (Figure 4.24).

Figure 4.24 A Newton pair

A Newton pair of forces has the following properties:
- The two forces act on two different bodies.
- Both forces are always of the same type (i.e. both gravitational, both electrostatic, etc.).
- The forces are equal in magnitude.
- The forces act in opposite directions.

A book on a table can be used to explain the idea of a Newton pair. In this example there are *two* Newton pairs:
- Gravitational forces — the pull of the Earth on the book and the pull of the book on the Earth.
- Contact forces — the push of the book on the table and the push of the table on the book.

> **Exam tip**
>
> Notice that in each case if one of the forces of the pair is removed it makes the other one vanish.

Now test yourself

15 A car of mass 1000 kg pulls a caravan of mass 800 kg. The driving wheels of the car exert a force of 8000 N on the road. The total resistance to motion is 3000 N.
  (a) What is the net accelerating force?
  (b) What is the acceleration?
  (c) What is the force of the car on the caravan?

Answer on p. 217

Momentum

Impulse

REVISED

> When a force acts on a body the velocity of the body may change. The product of the force and the time for which it acts is called the **impulse**.
>
> **impulse = force × time   (units N s)**

If application of an **impulse** is represented by a force–time graph, the impulse is the area beneath the line on the graph (Figures 4.25 and 4.26).

Momentum and momentum change

REVISED

> The change in the velocity of a body due to the action of an impulse depends not only on the size of the impulse but also the mass of the body. The product of the mass of the body and its velocity is called the **momentum** of the body.
>
> **momentum = mass × velocity   (units kg m s⁻¹)**
>
> Therefore, an impulse produces a change of momentum.

If the velocity of a body of mass m is changed from u to v by a force F acting for a time t:

impulse = Ft = momentum change (Δmv) = $mv - mu$ (units kg m s⁻¹ or N s)

Force may therefore be written as:

force = rate of change of momentum $F = \dfrac{\Delta mv}{\Delta t}$

> **Exam tip**
>
> Remember that momentum, and therefore momentum change, is a vector.

Example

Figure 4.25 shows an impulse of 8 N being applied to a body for 20 s.

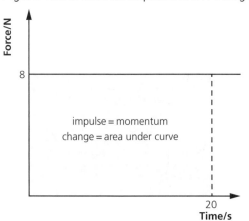

Figure 4.25 **Force–time graph**

Both the impulse and the momentum change are the area under the line. Therefore:

impulse = 8 × 20 = 160 N s

16 The graph in Figure 4.26 shows a varying force being applied to a body. If the body has a mass of 2.5 kg, calculate the impulse and hence the change in the velocity of the body during the first 10 s of the motion shown.

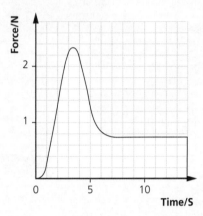

Figure 4.26 **Force–time graph, with variable force**

Answer on p. 217

Conservation of momentum

Momentum is conserved in a collision or explosion in an isolated system where no external forces act. In other words the momentum before the collision or explosion is the same as that after it. This is true for *all* collisions and explosions.

momentum before collision = momentum after collision

In a collision the same law of conservation of momentum applies. If a mass m_1 moving at a velocity u_1 collides with a mass m_2 moving at a velocity u_2 such that after the collision m_1 moves at v_1 and m_2 moves at v_2:

$m_1u_1 + m_2u_2 = m_1v_1 + m_2v_2$

The law of conservation of momentum applies whether the collisions are elastic or not (Figure 4.27).

Figure 4.27 Elastic collision

The special case of two equal masses making a completely elastic collision is shown in Figure 4.28. In this collision the velocities of A and B are swapped over.

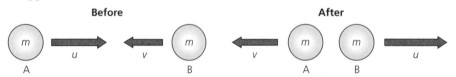

Figure 4.28 Elastic collision — equal masses

In a perfectly inelastic collision all the kinetic energy of the colliding bodies is lost — this may be converted into heat or used to deform the bodies. Imagine two balls colliding head on, sticking together and then moving off after the collision.

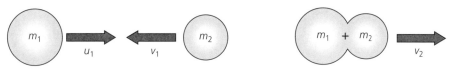

Figure 4.29 Perfectly inelastic collision

Explosions

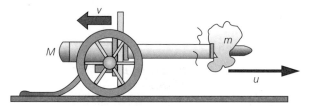

Figure 4.30 Cannon firing a shell

Energy and momentum are also conserved in explosions, although the type of energy may be changed. For example when a cannon fires a shell the total momentum of the shell (mu) (plus that of any exhaust gases etc.) and the cannon (Mv) is the same after firing as it was before firing — that is, zero.

> **Exam tip**
>
> In an explosion where two fragments are produced the heavier fragment will have the smaller velocity.

momentum before explosion = momentum after explosion

Therefore:

$0 = mu + Mv$

$Mv = -mu$

Example

If a mass of 3.5 kg moving left to right at $5\,m\,s^{-1}$ collides with a mass of 4.0 kg moving right to left at $3.0\,m\,s^{-1}$ and they stick together, find the final velocity of the combined masses.

Answer

momentum before impact = $(3.5 \times 5.0) + (-4.0 \times 3.0) = 5.5\,N\,s$

But this must equal the momentum after the collision, i.e. total mass × final velocity.

Notice that one of the velocities is negative, showing that the ball was moving right to left.

mass afterwards = 7.5 kg

Therefore:

velocity afterwards $= \dfrac{5.5}{7.5} = 0.73\,m\,s^{-1}$

This is positive, showing that after collision the combined two balls move from left to right.

Now test yourself   TESTED

17 A child throws a 200g snowball with a speed of 8 m s⁻¹ so that it hits the 1.5 kg head of a snowman. The snowball sticks to the snowman's head and knocks it off. What is the initial velocity of the ball and head just after collision?

Answer on p. 217

Work, energy and power

Work   REVISED

When a force moves an object **work** is done on the object and energy is converted from one form to another. The units for work are joule (J).

work done = force × displacement

= force × displacement in the direction of the force

work done = $F(s\cos\theta)$

> **Exam tip**
>
> Force and displacement are both vectors but work is a scalar.

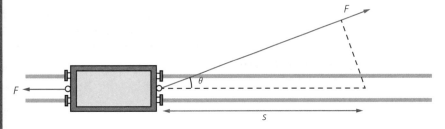

Figure 4.31 Work done by a force

Figure 4.31 shows a truck being pulled at a constant velocity a distance s along a pair of rails by a force F. The force is applied at an angle θ to the rails. The truck has a displacement s along the rails and the truck moves a distance $s\cos\theta$ in the direction of the force F.

If there were no friction between the rails and the truck, the force needed to keep the truck moving would be zero. However if there is a force of friction F' between the rails and the truck, once the truck is moving it will require a force $F\cos\theta$ (= F') acting left to right to keep the truck moving at a constant velocity.

> **Exam tip**
>
> In the extreme case of the force acting at right angles to the rails the truck would not move along the rails and the work done on the truck would be zero. Clearly the most effective direction in which to apply the force to the truck if we want it to move along the rails is parallel to the rails.

Example 1

Calculate the minimum work done to pull a truck 8 m along a pair of rails at constant velocity if the frictional force opposing the motion is 100 N.

Answer

In this case the minimum work done would be when the force is parallel to the rails.

work done = force × displacement = 100 × 8 = 800 J

Example 2

The force is now applied at an angle of 30° to the rails. If the frictional force remains the same, calculate:

(a) the force required to keep the truck moving at the same constant velocity along the rails

(b) the work done in moving the truck 8 m along the rails.

Answer

(a) force $(F)\cos\theta$ = frictional force = 100 N

(b) Therefore:

$$F = \frac{100}{\cos 30} = 115.5\,\text{N}$$

work done = 115.5 × 8 = 923.8 J

The work done on an object is the area under the line in a force–displacement graph. This applies if the force is constant (Figure 4.32(a)) or varying (Figure 4.32(b)).

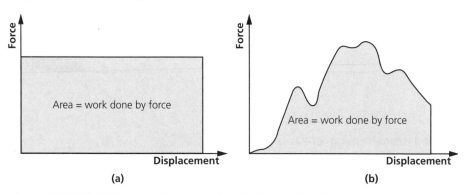

(a)   (b)

Figure 4.32 Work done and area under the force–displacement curve

work done against the force of gravity = $mg\Delta h$

where Δh is the change in height of the object above some original position measured upwards from that position.

Now test yourself

TESTED

18 A man pulls a boat along a canal at a constant speed using a rope that makes an angle of 30° with the direction that the boat moves along the canal. If the force in the rope is 500 N calculate:

(a) the component of force along the canal

(b) the component of the force at right angles to the canal

(c) the work done by the man to move the boat 40 m along the canal.

Answer on p. 217

Power

The rate at which work is done, or the rate at which energy is converted from one form to another is the **power**, and is defined as:

$$\text{power, } P = \frac{\text{work done}}{\text{time taken}} = \frac{\Delta W}{\Delta t}$$

The units of power are watts (W), where 1 watt is 1 joule per second.

> **Exam tip**
>
> Power can also be expressed as:
>
> **power = force (F) × velocity (v)**

Now test yourself

19 A car travelling at $30\,\text{m}\,\text{s}^{-1}$ along a level road is brought to rest in a distance of $35\,\text{m}$ by its brakes. If the total frictional drag at $30\,\text{m}\,\text{s}^{-1}$ is $5000\,\text{N}$ and the force exerted by the brakes during braking is $7000\,\text{N}$ calculate:
   (a) the power of the car when travelling at $30\,\text{m}\,\text{s}^{-1}$
   (b) the work done by the brakes to bring it to rest.

Answer on p. 217

Energy

Efficiency

Machines are devices for converting (transforming) one form of 'useful' energy to another form of 'useful' energy. How effective the machine is at making this transformation is called the **efficiency** of the machine.

$$\text{efficiency} = \frac{\text{useful energy transferred in a given time}}{\text{energy supplied in that time}} \times 100\%$$

A petrol engine is about 30% efficient, a diesel engine 40% efficient and our bodies are a mere 25% efficient — only one quarter of the energy produced goes to moving the muscles.

Conservation of energy

Principle of energy conservation

It is important to talk about the transformation or conversion of energy from one form to another and not its use. This is because although we may use up energy in one form it always reappears as another.

The principle of **conservation of energy** is that energy is never created or destroyed but only transformed from one form of energy to another.

Sources of energy

These include:
- fossil fuels — coal, oil, gas
- wind
- waves
- tides
- peat
- hydroelectric
- pumped storage
- nuclear fission
- nuclear fusion
- solar
- osmotic pressure
- geothermal
- biomass

Different forms of energy

REVISED

Energy can 'exist' in the following forms:
- mechanical (kinetic, gravitational potential)
- tensile
- nuclear
- heat (radiant, kinetic)
- magnetic
- sound (kinetic)
- electrical
- chemical
- mass

Gravitational potential energy

The energy associated with the position of a body of mass m in a gravitational field is the **gravitational potential energy** of the body compared with some reference point where $h = 0$ — usually the surface of the Earth.

If the distance moved parallel to the gravitational field is Δh then the change in potential energy is:

gravitational potential energy change = $mg\Delta h$

Exam tip

Remember that in gravitational potential energy changes it is the vertical height moved in the field that is important.

Example 1

A crane lifts a load of 300 kg through a distance of 2.5 m onto a truck. Calculate the gain in gravitational potential energy. (gravitational field strength = 9.8 N kg^{-1}).

Answer

gravitational potential energy gained = $mg\Delta h$ = 300 × 9.8 × 2.5 = 7350 J

Example 2

A mass of 25 kg is moved a distance of 35 m at an angle of 20° to a gravitational field of strength 9.8 N kg^{-1}. Calculate the change in gravitational energy.

Answer

change in gravitational potential energy = 25 × 9.8 × 35 cos 20 = 8058 J

Now test yourself

20 An 85 kg athlete trains by running up a flight of 30 steps. If each step is 15 cm high and 20 cm wide calculate the change in his gravitational potential energy. (gravitational field strength = 9.8 N kg^{-1})

Answer on p. 217

Kinetic energy

The energy possessed by a body by virtue of its motion is called the **kinetic energy** of the body.

The **kinetic energy** of an object depends on two things:
● the mass of the object (m)
● its speed (v)

The formula for kinetic energy of an object of mass m travelling at velocity v is:

$$\text{kinetic energy} = \tfrac{1}{2}mv^2$$

Kinetic energy changes

It is important to understand the correct way to calculate changes in the kinetic energy of an object. For example, suppose we want to find the increase in the kinetic energy of an 8 kg ball when its velocity is increased from 3 m s^{-1} to 4 m s^{-1}. The correct way is as follows:

kinetic energy increase (Δke) = $\tfrac{1}{2} \times 8 \times (4^2 - 3^2) = 4 \times (16 - 9) = 4 \times 7 = 28$ J

and *not*

kinetic energy increase = $\tfrac{1}{2} \times 8 \times (4 - 3)^2 = 4$ J

Exam tip

Check that you understand the correct way of calculating kinetic energy changes.

Example 1

A lorry of mass 6000 kg travels along a level road at 30 m s^{-1}. The brakes are then applied and the lorry stops in 70 m. Calculate:
(a) the kinetic energy of the lorry before braking
(b) the braking force.

Answer

(a) kinetic energy = $\tfrac{1}{2}mv^2 = \tfrac{1}{2} \times 6000 \times 30^2 = 2\,700\,000$ J = 2.7 MJ

(b) braking force = $\dfrac{\text{energy change}}{\text{braking distance}} = \dfrac{2.7 \times 10^6}{70} = 38.6$ kN

Exam practice answers and quick quizzes at **www.hoddereducation.co.uk/myrevisionnotes**

Example 2

A lorry of mass 2000 kg moving at $10\,m\,s^{-1}$ on a horizontal surface is brought to rest in a distance of 12.5 m by the brakes being applied.

(a) Calculate the average retarding force (F).

(b) What power must the engine produce if the lorry is to travel up a hill of 1 in 10 at a constant speed of $10\,m\,s^{-1}$ the frictional resistance being 200 N?

Answer

(a) kinetic energy of lorry = $\frac{1}{2} \times 2000 \times 100 = 10^5\,J = F \times 12.5$

Therefore $F = 8000\,N$

(b) On the hill, height risen per second = 1 m and distance travelled along the slope = 10 m.

potential energy gained by lorry per second (taking $g = 9.8\,N\,kg^{-1}$) = $2000 \times 9.8 \times 1 = 19\,600\,J$

work done against friction per second = $200 \times 10 = 2000\,J$

total energy required per second = $21\,600\,W = 21.6\,kW$

> **Exam tip**
>
> Remember that both gravitational energy and kinetic energy are scalars.

Now test yourself

TESTED

21 What is the maximum speed at which an earth-mover of mass 250 000 kg can descend a slope of 1 in 10 if the brakes can dissipate energy at a maximum rate of 2000 kW? ($g = 9.8\,N\,kg^{-1}$)

22 A lift has a mass of 400 kg. A man of mass 70 kg stands on a weighing machine fixed to the floor of the lift. Four seconds after starting from rest the lift has reached its maximum speed and has risen 5 m.
  (a) What will be the reading on the weighing machine during the period of acceleration?
  (b) How may it be decided whether the acceleration was uniform?
  (c) How much energy will be used by the lift motor in:
    (i) the first four seconds
    (ii) the next four seconds?

Answers on p. 217

Bulk properties of solids

Density

REVISED

The mass of individual atoms and how closely they are packed together can be 'felt' on an everyday level — it is called the **density** of the material.

$$\text{density, } \rho = \frac{\text{mass}}{\text{volume}} = \frac{m}{V}$$

Example

A statue has a volume of $5 \times 10^{-4}\,m^3$ and a mass of $4.75\,kg$. It has been made of copper (density $8930\,kg\,m^{-3}$) with a layer of silver (density $10\,500\,kg\,m^{-3}$) on top. What are the masses of copper and silver in the statue?

Answer

mass = volume × density

$4.75 = \rho_{Cu}V_{Cu} + \rho_{Ag}V_{Ag} = 8930\,V_{Cu} + 10\,500\,V_{Ag}$

But $V_{Cu} + V_{Ag} = 5 \times 10^{-4}$ and therefore $4.75 = 8930\,V_{Cu} + 10\,500(5 \times 10^{-4} - V_{Cu})$.

Therefore:

$V_{Cu} = \dfrac{0.5}{1570} = 3.18 \times 10^{-4}\,m^3$

So:

$V_{Ag} = 1.82 \times 10^{-4}\,m^3$

mass of copper = $8930 \times 3.18 \times 10^{-4} = 2.84\,kg$

mass of silver = $1.91\,kg$

> **Exam tip**
>
> Remember to use the correct SI units when calculating and expressing density.

Now test yourself

TESTED ☐

23 $3\,mg$ of gas are injected into the vacuum chamber of a fusion reactor. The volume of the chamber containing the gas is $3.75\,m^3$. What is the density of the gas under these conditions?

Answer on p. 217

Hooke's law and the elastic limit

REVISED ☐

The simplest form of variation of the extension of a metal wire when a stretching force is applied to it is known as **Hooke's law**. It relates the applied force (F), or load, to the increase in length, or extension (ΔL), of the object.

> **Hooke's law** states that if the elastic limit is not exceeded the extension is directly proportional to the applied force — doubling the force will double the extension.
>
> **force = constant × extension**
>
> $F = k\Delta L$
>
> The constant k is known as the **elastic constant** for the material and is defined as $F/\Delta L$. The units for k are $N\,m^{-1}$.

If a graph of force is plotted against extension a straight line will be obtained up to a certain point called the **elastic limit** (shown as P on the graph in Figure 4.33). Up to the elastic limit the wire behaves elastically — that is, it will return to its original length if the load is removed. Beyond the elastic limit the wire will remain permanently stretched.

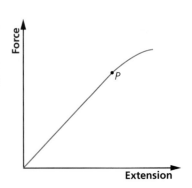

Figure 4.33 Hooke's law

TESTED

Example

An elastic cord has an unstretched length of 35 cm. One end is fixed to a support, and when a force of 2 N is applied to the lower end the length of the cord is 65 cm.

Calculate the elastic constant of the cord.

Answer

elastic constant, $k = \dfrac{\text{force}}{\text{extension}} = \dfrac{2}{0.3} = 6.7\,\text{N m}^{-1}$

Now test yourself

24 A load of 50 N is hung on a 2.50 m length of copper wire. The elastic constant for the wire is 9400 N m⁻¹. Calculate the new length of the wire.

Answer on p. 217

Tensile stress and tensile strain

REVISED

Tensile stress

Every elastic **stress** produces an elastic **strain**.

Tensile **stress** is a measure of the cause of the deformation produced by a force:

$$\text{tensile stress, } \sigma = \frac{\text{force}}{\text{area normal to the force}}$$

The units for tensile stress are N m⁻² or Pa.

Tensile strain

Strain is a measure of the deformation produced by the stress:

$$\text{tensile strain, } \varepsilon = \frac{\text{extension}}{\text{original length}}$$

Strain has no units as it is simply a ratio of two quantities with the same units.

Breaking stress

REVISED

The maximum stress that a material can stand before it breaks is called the breaking stress. There are two types of breaking stress:

- compressive breaking stress — the maximum squashing stress before fracture
- tensile breaking stress — the maximum stretching stress before fracture

The compressive breaking stress (*F*) of a material can be used to work out the maximum height of a rock column that is possible on the surface of the Earth (Figure 4.34).

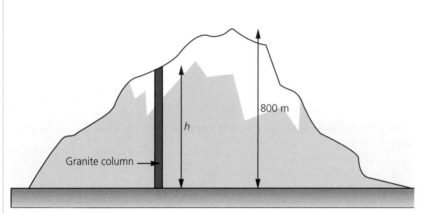

Figure 4.34 **Maximum height of a granite column**

maximum pressure at the base = $F = \rho g h$, where ρ is the density of the rock.

For granite:

$F = 145 \times 10^6$ and $\rho = 2500\,kg\,m^{-3}$

So:

$$h = \frac{145 \times 10^6}{2500 \times 9.81} = 5900\,m$$

This is smaller than the height of many mountains but in a mountain the 'column' of rock would be supported from the sides by the rest of the mountain.

Now test yourself

TESTED

25 A lift and its passengers have a combined a weight of 20 000 N. The lift is suspended from four steel cables of equal diameter. Calculate the minimum diameter of each cable, allowing for a 50% safety margin. (breaking stress for steel = 500 MPa)

Answer on p. 217

Elastic strain energy

REVISED

When a person jumps up and down on a trampoline it is clear that the bed of the trampoline stores energy when it is in a state of tension. This energy is converted to kinetic and gravitational potential energy of the jumper when the tension is removed.

Exam practice answers and quick quizzes at **www.hoddereducation.co.uk/myrevisionnotes**

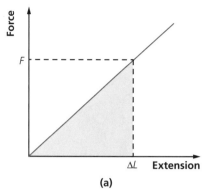 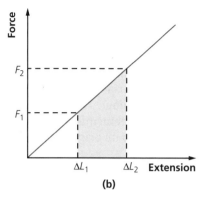

| (a) | (b) |

Figure 4.35 Elastic strain energy

Consider a wire of unstretched length L and let a force F produce an extension ΔL. Assume that the elastic limit of the wire has not been exceeded and that no energy is converted to other forms such as heat (Figure 4.35(a)).

The work done by a force is Fs but in this case the force varies from 0 at the start to F at the end when the wire is stretched by an amount ΔL. Therefore:

elastic energy stored in the wire = ½$F\Delta L$

But the work done by the force F is equal to the energy gained by the wire. Therefore:

work done on the wire during stretching = average force × extension
= ½$F\Delta L$

This energy is the shaded area of the graph.

If the extension is increased from ΔL_1 to ΔL_2 (Figure 4.35(b)) then the extra energy stored is given by:

additional elastic energy stored in the wire = ½$F(\Delta L_2 - \Delta L_1)$

Example

Calculate the energy stored in a stretched copper wire if its extension is increased by 1.5 mm when the force applied to it is increased by 50 N.

Answer

additional energy stored = ½$F(\Delta L_2 - \Delta L_1)$ = ½ × 50 × 1.5 × 10^{-3} = 0.0375 J

Now test yourself

TESTED

26 Calculate the increase in the elastic energy stored in the string of a cello if an increase in tension from 0 to 70 N produces an extension of 1 mm.

Answer on p. 218

Ductile, plastic and brittle materials

A **ductile** material is one such as copper, which may be drawn out into a wire without fracture. If a ductile material is stretched beyond its elastic limit it will show **plastic** behaviour. This means that when the load is removed some, or all, the deformation will be permanent and the material will not return to its original length before stretching.

Materials such as glass that can be extended but do not show plastic deformation and will easily fracture are known as **brittle** materials.

Stress–strain curves

If a ductile material such as copper is stretched it will follow the curve shown by Figure 4.36(a). From O to P Hooke's law is obeyed. The point E is the elastic limit — if the load is removed before E is reached the material will return to its original length. Between E and Y the material becomes plastic — not all the extension is recoverable if the load is removed. B is the breaking stress at which the material fractures.

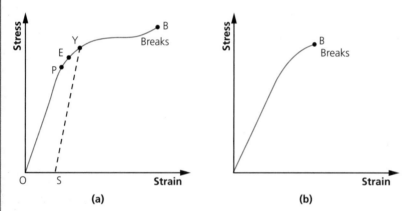

Figure 4.36 Ductile and brittle materials

Figure 4.36(b) shows the stress–strain curve for a brittle material. There is no plastic deformation.

The Young modulus

The **Young modulus** determines the relation between tensile stress and tensile strain (Figure 4.37).

Figure 4.37 The Young modulus

Exam practice answers and quick quizzes at **www.hoddereducation.co.uk/myrevisionnotes**

The **Young modulus** is the ratio of tensile stress to tensile strain:

$$\text{tensile stress } (\sigma) = \frac{\text{force}}{\text{cross sectional area}} = \frac{F}{A}$$

$$\text{tensile strain } (\varepsilon) = \frac{\text{extension}}{\text{original length}} = \frac{\Delta L}{L}$$

$$\text{Young modulus} = \frac{\text{tensile stress}}{\text{tensile strain}} = \frac{\sigma}{\varepsilon} = \frac{F/A}{\Delta L/L} = \frac{FL}{A\Delta L}$$

Example

A steel wire 10 m long and with a cross-sectional area of 0.01 cm² is hung from a support and a load of 150 N is applied to its lower end. Calculate the new length of the wire. (Young modulus for steel = 210 GPa)

Answer

$$\text{extension, } \Delta L = \frac{150 \times 10}{2.1 \times 10^{11} \times 1 \times 10^{-6}} = 7.14\,\text{mm}$$

Therefore, the new length = 10.00714 m

Now test yourself

TESTED

27 The rubber cord of a catapult has a cross-sectional area of 1.0 mm² and a total unstretched length of 10.0 cm. It is stretched to 15 cm and then released to project a missile of mass 5.0 g vertically. Calculate:
   (a) the energy stored in the rubber
   (b) the velocity of projection
   (c) the maximum height that the missile could reach.
   Take the Young modulus for rubber to be 5.0×10^8 Pa and $g = 9.8\,\text{N kg}^{-1}$.

28 It has been calculated that during running the force on the hip joint is about five times the body weight. Estimate the compression of the femur during each running stride for a sprinter of mass 70 kg. (Assume that the femur is 0.40 m long and has a mean diameter of 2.0 cm; $g = 9.8\,\text{N kg}^{-1}$; Young modulus for bone = 18×10^9 Pa.)

29 A gymnast of mass 70 kg hangs by one arm from a high bar. If the gymnast's whole weight is assumed to be taken by the humerus bone (in the upper arm) calculate the stress in the humerus if it has a radius of 1.5 cm. ($g = 9.8\,\text{N kg}^{-1}$)

Answers on p. 218

Required practical 4

Measurement of the Young modulus

The Young modulus can be measured for a material in the form of a wire using the apparatus shown in Figure 4.38.

Two identical wires are hung from a beam; one wire is used as a reference standard and has a scale is fixed to one wire and a mass hung on the end to remove kinks in it. The other wire has a small load placed on it to straighten it and a vernier scale that links with the scale on the reference wire.

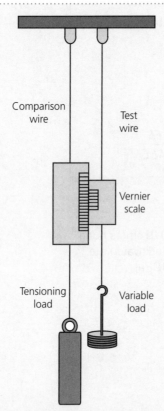

Figure 4.38 Measurement of the Young modulus

The original length (*L*) of the test wire is measured and its diameter is found for various points along its length.

Loads are then placed gently on the wire and the extension of the wire found for each one. They should not be dropped, as this would subject the wire to a sudden shock. After each reading the load should be removed to check that the wire returns to its original length, showing that its elastic limit has not been exceeded.

A graph is plotted of stress against strain and from this the value of the Young modulus can be found (this is the gradient of the line, i.e. *F/A* divided by Δ*L/L*).

Exam practice

Take *g* = 9.8 N kg⁻¹ where needed.

1. An oarsman rows a boat across a river that is flowing from left to right at 10 m s⁻¹. If the speed of the boat at right angles to the bank is 6 m s⁻¹ find the final velocity of the boat as it moves across the river. [4]
   You should use both the scale diagram and calculation methods to find your answers.

2. A child pulls a 120 kg sledge along a rough level road using a rope that is inclined at 35° to the horizontal. If the force in the rope is 300 N and the frictional force between the sledge and the road is 20 N what is the acceleration of the sledge? [3]

3. A stone is projected upwards at 25 m s⁻¹ at an angle of 30° to the horizontal (air resistance should be ignored for parts (a)–(d)):
   (a) What is its vertical velocity at the top of its flight? [1]
   (b) What is its horizontal velocity at the top of its flight? [2]
   (c) What is its horizontal velocity 5 s after it is fired? [2]
   (d) What is the maximum height it reaches? [3]
   (e) Sketch the trajectory of the projectile for the following two cases:
      (i) when air resistance is ignored [2]
      (ii) when air resistance is taken into account [3]

4 A rigid body acted on by a set of forces is in equilibrium if:
   A   the resultant force is zero
   B   the forces all act in the same direction and the resultant couple is zero
   C   the resultant force and the resultant couple are both zero
   D   the resultant couple is zero   [1]

5 A uniform ladder 3 m long and weighing 200 N leans against a wall so that it makes an angle
   of 55° with the ground. ($g = 9.8\,N\,kg^{-1}$)
   (a) What is the normal reaction between the ladder and the wall?   [3]
   (b) What is the size and direction of the reaction at the ground?   [3]
   (c) A painter of mass 70 kg climbs the ladder. Calculate the new values for parts (a) and
        (b) when they are 1.0 m up the ladder (measured along the ladder itself).   [4]

6 An engine of mass 5000 kg pulls a train of ten trucks each of mass 2000 kg along a horizontal track.
   Assume the frictional forces amount to 5000 N and that the engine exerts a force of 50 000 N on the
   rails. If the trucks are numbered from 1 to 10 starting with the one next to the engine, calculate:
   (a) the net total accelerating force   [2]
   (b) the acceleration of the train   [1]
   (c) the force of truck 6 on truck 7   [2]
   (d) the force of truck 9 on truck 8   [2]

7 A heavy ball with a mass of 5 kg is thrown with a velocity of 6 m s^{-1} to a boy who is standing on a
   skateboard at rest. He catches it and as a result moves backwards at 0.5 m s^{-1}. What is the
   combined mass (m) of the boy and skateboard?   [3]

8 Two bodies, P and Q, of equal mass move towards each other at speeds u and v respectively.
   They make an elastic collision, and during the collision P is momentarily at rest. What is the
   speed of Q at that moment?
   A $v - u$
   B $2(v - u)$
   C   zero
   D $\sqrt{uv}$ [1]

9 A firework rocket moves upwards and then explodes into two unequal fragments at the top of its
   flight path. They both move horizontally immediately after the explosion, one moves to the left at
   25 m s^{-1} and the other to the right at 75 m s^{-1}.
   (a) If the mass of the fragment moving to the left is 800 g what is the mass of the other fragment? [3]
   (b) What is the kinetic energy of the fragment moving left immediately after the explosion?   [2]
   (c) What is the kinetic energy of the other fragment immediately after the explosion?   [2]
   (d) Why is it necessary that the two fragments initially move off in exactly opposite directions?   [2]

10 A car of mass m has an engine that can produce a power P. What is the minimum time in which the
   car can be accelerated from rest to a speed v?
   A $\dfrac{mv^2}{P}$
   B $\dfrac{2P}{mv^2}$
   C $\dfrac{mv^2}{2P}$
   D $\dfrac{mv^2}{4P}$ [1]

11 A load of 50 kg is lifted by a small crane that is 35% efficient. If the load rises 6 m in 3 seconds,
   calculate the power used by the crane.   [2]

12 A monofilament nylon fishing line of original length 1.5 m and diameter 0.75 mm extends by 4 cm
   when a certain load is applied. (Young modulus for the fishing line = 6.5×10^9 Pa)
   (a) Calculate the elastic energy stored in the line.   [2]
   (b) The fishing line mentioned in part (a) will support a load of 100 N if it is applied steadily but will
        break when the same load is applied to it sharply. Why is this?   [2]

13 A steel wire of diameter d has a strain of 12.0×10^{-4} when supporting a certain load. If the wire is replaced by a second wire of the same material, but with a diameter of $d/2$, what will be the strain in this wire if it supports the same load?

A 6.0×10^{-3}

B 4.8×10^{-3}

C 2.4×10^{-3}

D 6.0×10^{-4} [1]

14 The gravitational field strength at the surface of a neutron star is $1.35 \times 10^{12}\,N\,kg^{-1}$. What would be the theoretical maximum height of a cylindrical granite column that could support its own weight without crushing when exposed to a field of this magnitude? (density of granite = $2700\,kg\,m^{-3}$; crushing strength = $3.6 \times 10^{6}\,Pa$)   [2]

Answers and quick quiz 4 online

ONLINE

Summary

You should now have an understanding of:

- scalars and vectors — scalar quantities have only magnitude, while vector quantities have both magnitude and direction
- moments — the turning effect of a force about a point; moment = force × perpendicular distance from the point to the line of action of the force; for a body to be in equilibrium, both the resultant force and the resultant moment must be zero
- motion along a straight line — governed by the equations of motion for uniform acceleration
- projectile motion — this can be considered in two parts, one horizontal (uniform motion) and one vertical (accelerated motion)
- Newton's laws of motion — the first law governs the motion of bodies under no resultant force; the second is concerned

with their behaviour with a resultant force; and the third explains the action of forces on two bodies
- momentum — mass × velocity; momentum is conserved in all collisions
- work, power (the rate at which work is done), energy and efficiency
- conservation of energy — energy is not created or destroyed but can be changed from one form to another
- density — mass/volume
- Hooke's law — the force is directly proportional to the extension up to the elastic limit
- the Young modulus — $(F/A)/(\Delta L/L)$; this is the modulus of elasticity that governs the linear extension of a specimen when a force is applied

5 Electricity

Basics of electricity

Electric charge

REVISED

When an electric current flows, electrical energy is converted to other forms of energy such as heat, light, chemical, magnetic and so on.

In a metal there is a large number of electrons that are not held around particular nuclei but are free to move at high speed and in a random way through the metal. These are known are **free electrons** and in a metal there are always large numbers of these. It is when these free electrons are all made to move in a certain direction by the application of a voltage across the metal that we have an electric current (Figure 5.1).

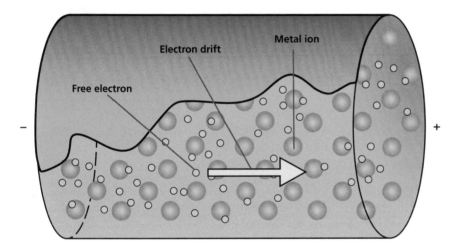

Figure 5.1 Free electrons in a wire

Each electron has only a very small amount of electric charge (e), so a larger unit is used when measuring practical units of charge. This unit is the **coulomb**:

1 coulomb = $-6.25 \times 10^{18}\,e$

Therefore the charge on one electron is $-1.6 \times 10^{-19}\,\text{C}$.

Electric current as a rate of flow of charge

REVISED

A **current** of 1A flows in a wire if a charge of 1C passes any point in the wire each second.

The rate of flow of electric **charge** ($\Delta Q/\Delta t$) round a circuit is the electric current in that circuit.

Typical mistake

Not using seconds for the time when calculating current.

Example 1

Calculate the current in a wire if a charge of 240 C passes a given point in 3 minutes.

Answer

$$\text{current} = \frac{Q}{t} = \frac{240}{180} = 1.33\,\text{A}$$

Example 2

A current of 25 mA flows for 10 ms.
(a) What charge has passed?
(b) How many electrons have flowed past that point in the circuit.

Answer

(a) charge = current × time = 0.025 × 0.010 = 0.00025 C = 2.5×10^{-4} C

(b) number of electrons $= \dfrac{2.5 \times 10^{-4}}{1.6 \times 10^{-19}} = 1.56 \times 10^{15}$

Now test yourself

TESTED

1 (a) A charge of 20 C passes a point in a circuit in 4 s. What is the current in the circuit? (Give your answer in amps.)
   (b) A charge of 600 C passes a point in a circuit in 20 minutes. What is the current in the circuit? (Give your answer in milliamps.)
2 (a) A current of 2 A flows for 10 s. What charge has passed? (Give your answer in coulombs.)
   (b) A current of 5 mA flows for 8 minutes. What charge has passed? (Give your answer in coulombs.)

Answers on p. 218

Note that we are using the conventional direction for electric current flow, i.e from positive to negative. In actual fact, of course, the electrons in a wire move from negative to positive when a current flows.

Resistance

REVISED

As the free electrons in an electric current move through the metal they collide with each other and with the atoms of the metal. These collisions impede their movement and this property of the material is called its **resistance**.

The **resistance** (R) of a given piece of material is connected to the current flowing through it (I) and the potential difference (V) between its ends by the equation:

$$\text{resistance} = \frac{\text{potential difference}}{\text{current}}$$

$$R = \frac{V}{I}$$

Exam practice answers and quick quizzes at **www.hoddereducation.co.uk/myrevisionnotes**

Potential difference

REVISED

As a charge moves round a circuit from the positive to the negative it loses energy.

> **The electric potential energy of a unit charge at a point in a circuit is called the potential at that point.**

The difference in electric potential between two points in the circuit is known as the **potential difference** (p.d.) between those two places.

> **Potential difference** between two points in a circuit is the work done (W) in moving unit charge (Q) (i.e. 1 coulomb) from one point to the other:
>
> $$\text{potential difference, } V = \frac{\Delta W}{\Delta Q}$$
>
> The units for both potential and potential difference are joules per coulomb, or volts (1 volt = 1 joule per coulomb).

Current–voltage characteristics

Ohm's law

REVISED

If the ratio of V to I remains constant for a series of different potential differences the material is said to obey Ohm's law and is known as an **ohmic conductor**.

This means that although we can always work out the resistance of a sample, knowing the current through it and the p.d. across it, if these quantities are altered we can only *predict* how it will behave under these new conditions if it obeys **Ohm's law**.

> **Ohm's law** states that the current in a conductor is directly proportional to the potential difference across it.

Figure 5.2 shows the variation of current (I) with potential difference (V) for a material that obeys Ohm's law — in other words, an ohmic conductor.

It is important to realise that Ohm's Law only holds for a *metallic conductor* at a *constant temperature*.

Figure 5.3 shows the variation in the potential around the circuit. We can follow this by considering each section of the circuit in turn.
- Along the connecting wire from the cell to B there is no resistance and so no loss of electrical energy or drop in potential.
- In the resistors r and R energy is converted to heat and so the potential drops from B through to E.
- From E to the cell there is no loss of electrical energy and so the potential at E is the same as that at the negative terminal of the cell — zero.

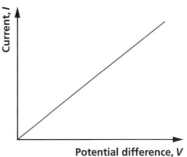

Figure 5.2 Ohm's law graph

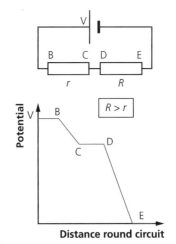

Figure 5.3 Variation in potential round a circuit

Example 1

A 6 V battery is connected to a small electromagnet and a current of 1.5 A flows through it. What is the resistance of the electromagnet?

Answer

$$\text{resistance} = \frac{\text{voltage}}{\text{current}} = \frac{6}{1.5} = 4\,\Omega$$

A current of 0.5 mA flows through a resistor of 100 kΩ. What is the potential difference across the resistor?

Answer

voltage = current × resistance = 0.0005 × 100 000 = 50 V

Now test yourself

3 Calculate the current through the following resistors:
  (a) 120 Ω connected to 200 mV
  (b) 4700 Ω connected to 12 V
  (c) 10 kΩ connected to 6 V
  (d) 2.5 MΩ connected to 25 V
4 What is the resistance of the following?
  (a) a torch bulb that draws 0.25 A from a 12 V supply
  (b) an immersion heater that draws 10 A from a 230 V supply

Answers on p. 218

Common current–voltage characteristics

A current sensor and a voltage sensor can be used to capture data to give V–I curves. Figure 5.4 shows some common examples.

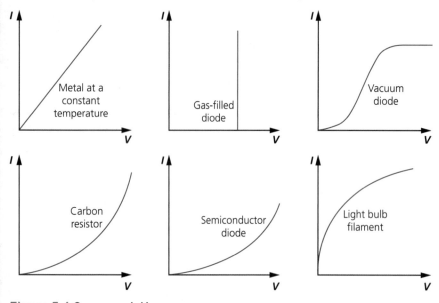

Figure 5.4 Common I–V curves

The graph for a metal at a constant temperature is an example of an ohmic conductor (see below).

Resistivity

What is resistivity?

There are three factors that affect the resistance of a sample of a material:
● the temperature
● the dimensions of the sample — the smaller the cross-sectional area and the longer the sample the larger the resistance
● the material from which the sample is made

The property of the material that affects its resistance is called the **resistivity** of the material (Figure 5.5), symbol ρ.

> The **resistivity** of a material is defined as the resistance between two opposite faces of a metre cube of the material. It is related to the resistance (R) of a specimen of length L and cross-sectional area A by the formula:
>
> $$\text{resistivity, } \rho = \frac{RA}{L}$$
>
> The units for resistivity are $\Omega\,\text{m}$.

Typical mistake

Using $\Omega\,\text{m}^{-1}$ instead of $\Omega\,\text{m}$ as the unit for resistivity.

Resistance, R

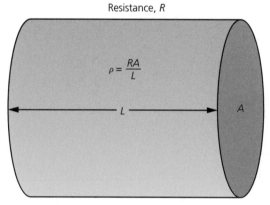

$$\rho = \frac{RA}{L}$$

Figure 5.5 Resistivity definition

The resistivities of good conductors are very small numbers (usually between 10^{-8} and $10^{-6}\,\Omega\,\text{m}$), rising to around $1\,\Omega\,\text{m}$ for semiconductors and $10^{15}\,\Omega\,\text{m}$ for 'insulators'.

Example 1

Calculate the resistance of a 1.5 m long piece of wire of resistivity $30 \times 10^{-8}\,\Omega\,\text{m}$ and diameter 0.5 mm.

Answer

$$\text{resistance} = \frac{\text{resistivity} \times \text{length}}{\text{area}} = \frac{30 \times 10^{-8} \times 1.5}{1.96 \times 10^{-7}} = 2.3\,\Omega$$

Example 2

Some resistance wire (resistivity $40 \times 10^{-8}\,\Omega\,\text{m}$) is used to make a heater. The wire on the reel has a cross-sectional area of $1.5 \times 10^{-7}\,\text{m}^2$, and the required resistance is $5\,\Omega$. What length of wire is needed?

Answer

$$\text{length} = \frac{\text{resistance} \times \text{area}}{\text{resistivity}} = \frac{5 \times 1.5 \times 10^{-7}}{40 \times 10^{-8}} = 1.88\,\text{m}$$

Exam tip

In resistivity calculations make sure that you use metres and not cm or mm, and radius and not diameter.

Required practical 5

Measurement of resistivity

The resistivity of a wire can be measured using a low-voltage power supply, a micrometer, an ammeter and a voltmeter.

- First measure the diameter of the wire in a number of places using the micrometer and calculate an average value.
- Then connect the wire and meters to the power supply and apply a small voltage.
- Take readings of the current through the wire and the potential difference across it.
- Hence calculate the resistivity.

It is important to avoid heating the wire by using too large a potential difference.

An alternative method uses an ohm meter instead of the ammeter, voltmeter and power supply.

Now test yourself

TESTED

5 Calculate the resistivity of a material if a 250 cm length of wire of that material with a diameter of 0.56 mm has a resistance of 3 Ω.

6 Calculate the resistance between the large faces of a slab of germanium of thickness 1 mm and area 1.5 mm². The resistivity of germanium is 0.65 Ωm.

Answers on p. 218

Resistance and temperature

REVISED

When a material is heated its resistivity will change and therefore so will the resistance of a specimen of that material. The nature of the change depends on the material. The change is governed by a property called the **temperature coefficient of resistance** (α). This is positive for metals but negative for non–metals such as semiconductors.

Metals

For a metal an increase in the temperature gives an increase in resistance. At low temperatures the thermal vibration of the lattice ions is small and electrons can move easily, but at high temperatures the motion increases, giving a much greater chance of collisions between the conduction electrons and the lattice ions, so impeding their motion.

The variation is shown in Figure 5.6.

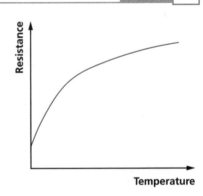

Figure 5.6 Metal resistance and temperature variation

Exam practice answers and quick quizzes at **www.hoddereducation.co.uk/myrevisionnotes**

Semiconductors

In semiconductors an increase in temperature leads to a drop in resistance. Bound electrons gain energy and move into the conduction band, resulting in an increase in the number of free electrons. The temperature coefficient of resistance is therefore negative. Such materials are called negative temperature coefficient (NTC) semiconductors.

Thermistors

The change of resistance of a semiconductor with temperature is used in temperature sensitive resistors called thermistors. The most widely used are NTC thermistors whose resistance falls as the temperature rises. The symbol for a thermistor and a graph of the variation of its resistance with temperature are shown in Figure 5.7.

Thermistors are used as temperature sensors in thermostats in ovens and irons, in fire alarms and on the wing of a plane to detect when the temperature falls low enough for ice to form. They are also in use in premature baby units to detect when a baby may have stopped breathing, current-limiting devices and thermometers.

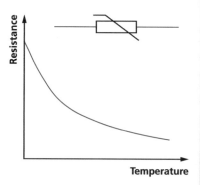

Figure 5.7 Resistance variation for an NTC thermistor

Now test yourself

TESTED

7 When is the filament in a 'traditional' light bulb most likely to break? Explain your answer.

Answer on p. 218

Superconductivity

REVISED

When metals cool, their resistance falls steadily as the motion of the atoms of the metal and the free electrons gets less and so the number of electron–atom collisions is reduced.

However, it was found that as the metal is cooled further a temperature can be reached where the resistance suddenly falls to zero — when this happens the metal is said to be **superconducting** and the phenomenon is called **superconductivity**.

The temperature at which this happens for a given metal is called the **critical temperature** for the metal.

The importance of superconductivity is that if a material is superconducting it has no resistance, this means that an electric current can flow through it without energy loss in the form of resistive heating.

Applications of superconductivity include:
- high-power superconducting electromagnets for use in both the levitation of experimental trains and in nuclear accelerators
- superconducting power cables for electrical energy transmission

Circuits

Resistors in series

A series circuit is one where the components are connected one after the other. This means that the current passing through all the components is the same.

In Figure 5.8 the current through both resistors is I and the potential difference across R_1 is V_1 and that across R_2 is V_2.

The total resistance (R) of a set of resistors in series is simply found by adding the values of the resistance of each resistor together:

$$R = R_1 + R_2 + R_3 + \ldots$$

The above formula is true no matter how many resistors you add.

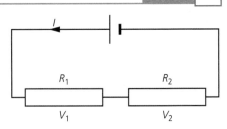

Figure 5.8 Resistors in series

Resistors in parallel

When the resistors are connected in parallel the current splits at the junction, a current I_1 passing through R_1 and a current I_2 passing through R_2. The potential difference across any number of resistances connected in parallel is the same for all the resistors (Figure 5.9).

The formula for total resistance (R) for resistors connected in parallel is:

$$\frac{1}{R} = \frac{1}{R_1} + \frac{1}{R_2} + \frac{1}{R_3} + \ldots$$

This version of the formula is true no matter how many resistors you add. However, a simpler version can be derived for two resistors in parallel:

$$R = \frac{R_1 R_2}{R_1 + R_2}$$

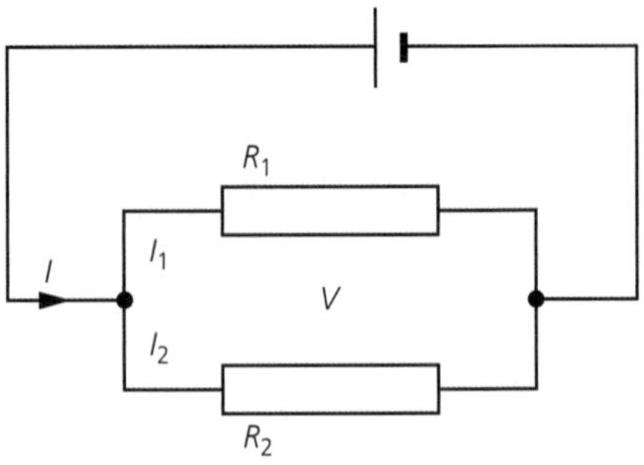

Figure 5.9 Resistors in parallel

> **Exam tip**
>
> The version in this form is only correct for two resistors.

> **Typical mistake**
>
> When calculating the final resistance for a pair of resistors in parallel, working out 1/R and then forgetting to invert it to obtain the final resistance R.

Example 1

Calculate the resistance of the following combinations:
(a) $200\,\Omega$ and $100\,\Omega$ in series
(b) $200\,\Omega$ and $100\,\Omega$ in parallel.

Answer

(a) $R = R_1 + R_2 = 200 + 100 = 300\,\Omega$

(b) $\dfrac{1}{R} = \dfrac{1}{R_1} + \dfrac{1}{R_2} = \dfrac{1}{200} + \dfrac{1}{100} = \dfrac{3}{200}$ and so $R = 67\,\Omega$

Example 2

You are given one $100\,\Omega$ resistor and two $50\,\Omega$ resistors. How would you connect any combination of them to give a combined resistance of:
(a) $200\,\Omega$
(b) $125\,\Omega$?

Answer

(a) $100\,\Omega$ in series with both the $50\,\Omega$
(b) the two $50\,\Omega$ in parallel and this in series with the $100\,\Omega$

Now test yourself

8 What is the final resistance of each of the six circuits in Figure 5.10?

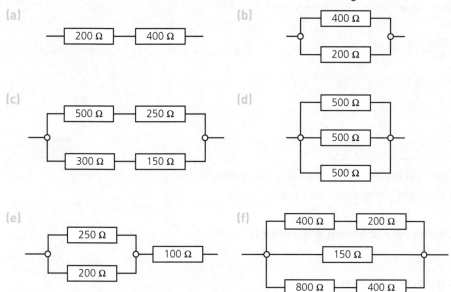

(a)

(b)

(c)

(d)

(e)

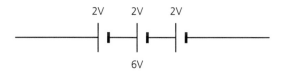

(f)

Figure 5.10

9 Figure 5.11 shows an LDR connected in parallel with a resistor and a 6 V cell of negligible internal resistance. The resistance of the LDR falls as the intensity of light falling on it is increased.

(a) Calculate the current flowing from the cell when the resistance of the LDR is 150 kΩ.

(b) What happens to this current if the light intensity is reduced?

(c) What is the minimum current that can be drawn from the cell?

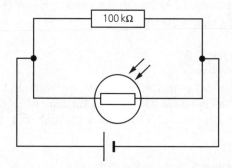

Figure 5.11 **LDR and resistor in parallel**

Answers on p. 218

Cells in series and parallel

REVISED ☐

For cells connected in series (Figure 5.12), the total potential difference between the ends of the chain is the sum of the potential differences across each cell.

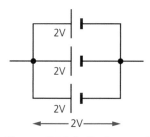

Figure 5.12 **Cells in series**

Figure 5.13 **Cells in parallel**

For cells connected in parallel (Figure 5.13), the total potential difference across the arrangement is the same as for one cell.

The advantage of the parallel circuit is that although the output voltage is the same as that of a single cell the battery formed from the group of cells contains more energy and so will supply current for longer.

Typical mistake

Not allowing for the change in total output voltage if one or more of the cells in a series is reversed.

Now test yourself

TESTED

10 You are given two 6V cells of negligible internal resistance and a 200 Ω resistor. What current flows through the resistor when the cells are connected
(a) to the resistor in parallel
(b) to the resistor in series?

Answer on p. 218

Electrical energy

REVISED

If a charge Q moves between two points in a circuit that have a potential difference of V volts between them the energy gained (or lost) by the charge is given by the formula:

electrical energy = charge (Q) × potential difference (V)

Since charge (Q) = current (I) × time (t):

electrical energy = IVt

Large amounts of energy are used in a car starter motor to 'turn the engine over'. Although the voltage is low (12V) the current required may be a great as 200A.

Example

Calculate the amount of energy supplied by a 6V battery when:
(a) a charge of 25C passes through it
(b) a current of 30mA flows through it for 5 minutes

Answer
(a) energy = potential difference × charge = 6 × 25 = 150 J
(b) energy = potential difference × charge
     = potential difference × current × time
     = 6 × 30 × 10^{-3} × 300 = 54 J

Typical mistake

Forgetting to convert to SI units, for example mA to A and minutes to seconds.

Now test yourself

TESTED

11 Calculate how much electrical energy is supplied by a 1.5V battery when:
(a) a charge of 3000C passes through it
(b) a current of 200 µA flows from it for 2.5 hours
12 How much energy is drawn from a 12V car battery if it is used to supply 200A for 1.5s to the starter motor?

Answers on p. 218

Power is the rate at which work is done or energy changed from one form to another, and so:

$$\text{electrical power} = \frac{\text{energy}}{\text{time}} = \frac{VQ}{t} = VI$$

Electrical power is measured in watts (W) where $1\,\text{W} = 1\,\text{J s}^{-1}$. For large power outputs we use kilowatts ($1\,\text{kW} = 1000\,\text{W}$) and megawatts ($1\,\text{MW} = 1\,000\,000\,\text{W}$).

Since $V = IR$, and power $= VI$:

$$\text{electrical power} = VI = I^2 R = \frac{V^2}{R}$$

Example 1

Calculate the current used by a 12 V immersion heater that is designed to deliver 30 000 J in 5 minutes.

Answer

energy = power × time = 30 000

Therefore:

30 000 = power × 300

power = 100 W

So:

$$\text{current} = \frac{100}{12} = 8.3\,\text{A}$$

Example 2

(a) Calculate the resistance of a 100 W light bulb if it takes a current of 0.8 A.
(b) Calculate the power of a 12 V immersion heater with a resistance of 10 Ω.

Answer

(a) power = $I^2 R$
Therefore $R = \dfrac{100}{0.64} = 156.3\,\Omega$

(b) power $= \dfrac{V^2}{R} = \dfrac{144}{10} = 14.4\,\text{W}$

Now test yourself

TESTED

13 What power is supplied to the heater of an electric bar fire with a resistance of 50 Ω connected to the mains 230 V supply?
14 What is the power loss down a copper connecting lead 75 cm long with a resistance of 0.13 Ω per metre when a current of 4.5 A flows through it?

Answer on p. 218

Conservation of charge and energy

In an electrical circuit both charge and energy must be conserved. These requirements are usually expressed in Kirchhoff's two rules:

1 The algebraic sum of the currents at a junction is zero. In other words there is no build up of charge at a junction ($\Sigma I = 0$).

2 The sum of the changes in potential round a closed circuit must be zero.

Rule 1 is about charge conservation while rule 2 is about energy conservation (Figure 5.14).

Rule 1 — at point B there is a junction:

current flowing from the cell (I) = Current in R_1 (I_1) + current in R_2 (I_2)

Rule 2 — round loop A–B–C–D–E:

p.d. across cell = − p.d. across R_1

This represents a gain of potential in the cell but a loss in R_1.

In this equation there is a minus because we are moving 'against' the current in R_2.

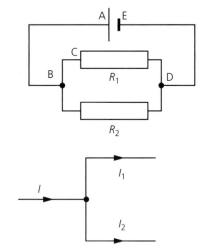

Figure 5.14 **Conservation of charge and energy**

Example

Consider the circuit in Figure 5.15.

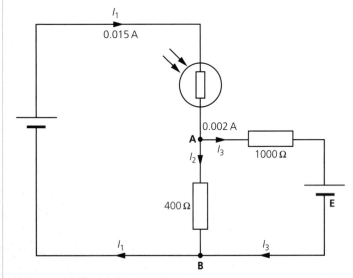

Figure 5.15 **Example problem circuit**

Applying Kirchhoff's first rule to junction A:

current in the 400 Ω resistor = 0.015 − 0.002 = 0.013 A

potential difference across 400 Ω resistor = 0.013 × 400 = 5.2 V

This is the potential difference between A and B via the 400 Ω resistor but it is also the potential difference across the right-hand branch of the circuit via the cell of emf ε (p. 109).

The potential drop across the 1000 Ω resistor is 0.002 × 1000 = 2 V.

Applying Kirchhoff's second rule to the right-hand branch and considering an anticlockwise direction from the cell:

EMF of the cell, E = (−0.002 x 1000) + 5.2 = −2 + 5.2 = 3.2 V

The minus sign is there because the current in the 1000 Ω resistor is travelling in the opposite direction from that in which the emf of the cell is acting.

Exam practice answers and quick quizzes at **www.hoddereducation.co.uk/myrevisionnotes**

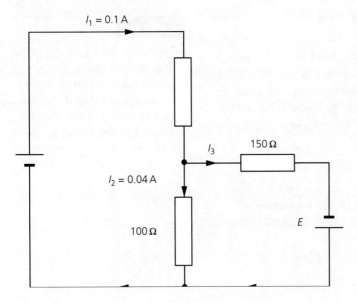

Figure 5.16

15 Using Figure 5.16:
   (a) Find the magnitude and direction of I_3.
   (b) Find the magnitude and direction of E.
16 Using Figure 5.16:
   (a) If $I_1 = 0.2\,A$ and $I_2 = 0.5\,A$, find the magnitude and direction of I_3.
   (b) If $I_1 = 0.3\,A$ and $I_2 = 0.1\,A$, find the magnitude and direction of I_3.

Answers on p. 218

Potential dividers

Basic circuit

REVISED ☐

Two resistors connected across a cell enable the output of the cell to be divided between them. Such a circuit is called a potential divider. The basic circuit is shown in Figure 5.17. If the output is continuously variable from 0 to V the device is known as a potentiometer. The p.d. across R_1 and R_2 is fixed (V). The output voltage across AB is given by:

$$\text{output voltage, } V_2 = \frac{R_2}{(R_1 + R_2)}V$$

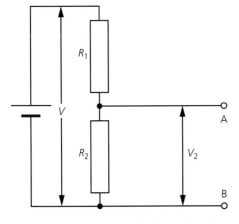

Figure 5.17 **Potential divider circuit**

Typical mistake

Taking the ratio of the two resistors rather than the ratio of one resistor to the sum of both.

Measuring the output voltage with a meter

This can be done using a **digital voltmeter** with very high (if not virtually infinite) resistance. The output voltage measured by this meter is that across R_2, in other words V_2.

Another option is to use a **moving coil meter**. These meters have a much lower resistance than a digital meter, usually some tens of kΩ. This means that the combined resistance of R_2 and the moving coil meter in parallel with it is less than R_2. The proportion of the input voltage (V) dropped across R_2 therefore falls and so the output voltage is less than that measured with a digital meter.

Exam tip

The total voltage across both the resistor and the other component in the circuit must always stay the same and be equal to the supply voltage of the battery.

Example

A loudspeaker is connected across the output (R_2) of a potential divider. Varying R_1 will change the potential across R_2 and so the device acts as a volume control (Figure 5.18).

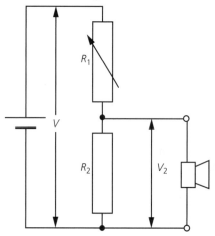

Figure 5.18 **Potential divider with loudspeaker**

If $V = 6\,V$, $R_1 = 200\,\Omega$, $R_2 = 0.5\,k\Omega$ and the loudspeaker has a resistance of $100\,\Omega$, calculate the p.d. across it.

Answer

Call the combined resistance of R_2 and the loudspeaker R_3:

$$R_3 = \frac{500 \times 100}{600} = 83\,\Omega$$

(This the resistance of R_2 (500 Ω) and the loudspeaker (100 Ω) in parallel.)

Therefore, using the formula for the potential divider:

$$V_2 = \frac{R_3}{(R_1 + R_3)} \times 6$$

p.d. across the speaker, $V_2 = (83/283)6 = 1.8\,V$

Light-dependent resistor (LDR)

An LDR is a component that has a resistance that changes when light falls on it. As the intensity of the light is increased, so the resistance of the LDR falls.

If the LDR is connected as part of a potential divider, as shown in Figure 5.19, then as the light level is increased its resistance falls and the proportion of the input voltage dropped across it will also fall.

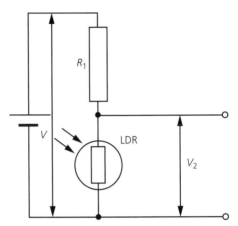

Figure 5.19 **Potential divider with LDR**

So in the light V_2 is low and in the dark V_2 is high.

Thermistor

If R_2 is replaced by an NTC thermistor the circuit is temperature dependent. As the temperature of the thermistor rises its resistance falls and so the voltage dropped across it falls.

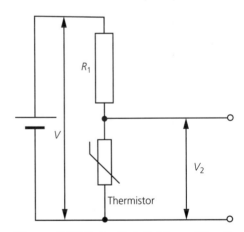

Figure 5.20 **Potential divider with a thermistor**

When the thermistor is hot V_2 is low and when the thermistor is cold V_2 is high.

Both these examples have considered R_2 being replaced by another component. If R_1 is replaced then if the voltage across this component rises the output voltage across R_2 will fall.

Revision activity

Make a mind map showing the various possible circuits using a potential divider (e.g resistors, thermistor and LDR). Summarise the effect on the output p.d. of changing the values of the components.

17 Using Figure 5.17, calculate the output voltage for the following
values of V, R_1 and R_2.
   (a) $V = 12\,V$, $R_1 = 100\,k\Omega$, $R_2 = 200\,k\Omega$
   (b) $V = 10\,V$, $R_1 = 25\,k\Omega$, $R_2 = 20\,k\Omega$
   (c) $V = 6\,V$, $R_1 = 250\,\Omega$, $R_2 = 200\,\Omega$
18 (a) Resistor R_1 is now replaced by a thermistor with a negative
   temperature coefficient — one where the resistance decreases
   as the temperature rises.
   If the values of the resistance of R_2 and the thermistor are
   equal at the start, what will happen to the output potential
   difference (V) as the thermistor is cooled?
   (b) Resistor R_2 is now replaced by a light-dependent resistor. (R_1 is
   a fixed resistor.)
   If the values of the resistance of R_1 and the LDR are equal
   at the start, what will happen to the output potential
   difference (V) as the intensity of the light falling on the LDR
   is decreased?

Answers on p. 218

The variable resistor as a potential divider　　　　REVISED ☐

Another way of varying the output potential difference is to use a variable
resistor or rheostat (Figure 5.21). This is made of a length wire wrapped
round a former.

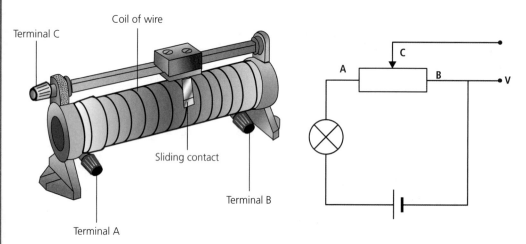

Figure 5.21 The variable resistor

The source potential is connected between A and B and the output (V)
taken between B and C.

Electromotive force and internal resistance

Electromotive force

When a charge passes through a cell it gains energy. The energy gained per coulomb in the cell is called the **electromotive force (emf)** (ε). It is the energy delivered per coulomb by the cell and so:

$$\text{electromotive force } (\varepsilon) = \frac{E}{Q}$$

When a current flows from the cell energy may be converted to other forms within the cell and the potential difference (V) between the terminals of the cell will then be less than the emf of the cell.

Note that emf is not a force. It is energy per unit charge, in other words a voltage.

> The **emf** (ε) of the cell is the maximum potential difference that the cell can produce across its terminals, or the open circuit potential difference.

Internal resistance

All cells have a resistance of their own and we call this the **internal resistance** (r) of the cell. The loss of electrical energy within the cell and the resulting reduction in the output potential difference is due to this internal resistance.

The internal resistance is related to the emf by the following equation:

$$\varepsilon = \frac{E}{Q} = V + Ir = I(R + r)$$

where I is the current flowing through the cell.

Figure 5.22 explains the ideas of emf and internal resistance.

The shaded area represents the internal part of the cell.

The quantity of useful electrical energy available outside the cell is IR and Ir is the energy transformed to other forms within the cell itself.

We usually require the internal resistance of a cell to be small to reduce the electrical energy transformed within the cell. The low internal resistance of a car battery allows it to deliver large currents without a large amount of electrical energy being converted to other forms within the battery itself. However it is sometimes helpful to have a rather larger internal resistance to prevent large currents from flowing if the cell terminals are shorted.

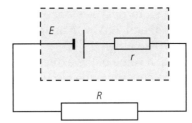

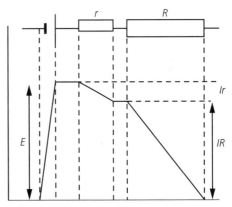

Figure 5.22 Emf and internal resistance

Example

A cell of emf 12 V and internal resistance 0.2 Ω is used in two circuits. Calculate the potential difference between its terminals when it is connected to:

(a) 15 Ω

(b) 0.1 Ω.

Answer

(a) total resistance = 15 + 0.2 = 15.2 Ω

Therefore:

$$current = \frac{12}{15.2} = 0.789\,A$$

'loss' of energy per coulomb in the cell = 0.789 × 0.2 = 0.158 V

potential difference between terminals = 12 − 0.158 = 11.84 V

(b) total resistance = 0.2 + 0.1 = 0.3 Ω

Therefore:

$$current = \frac{12}{0.3} = 40\,A$$

'loss' of energy per coulomb in the cell = 40 × 0.1 = 4 V

potential difference between terminals = 12 − 4 = 8 V

> **Exam tip**
>
> The word 'loss' is used here, although it should really be replaced with 'electrical energy converted to other forms'.

Now test yourself

TESTED

19 Explain what happens to the output potential across the terminals of a cell with some internal resistance as the current from it is increased.

20 A digital voltmeter with resistance of 10 MΩ reads 1.30 V when connected across the terminals of a cell. When the same meter is connected across a resistor of 20 Ω that has been connected in series with the cell the voltmeter reads 1.25 V.

Explain the difference between these two readings and calculate:

(a) the current in the external resistor

(b) the internal resistance of the cell.

21 A cell of emf 2.5 V with an internal resistance of 0.15 Ω is connected in turn to external resistors of

(a) 20 Ω and then (b) 1 Ω.

For each value of the external resistor calculate:

(i) the current in the circuit

(ii) the potential difference across the terminals of the cell

(iii) the power loss inside the cell.

Answers on pp. 218–9

Required practical 6

Investigation of the emf and internal resistance of electric cells

The emf (ε) of a cell can be measured using a high-resistance voltmeter connected between its terminals. The high resistance means that there is effectively zero current being drawn from the cell. The internal resistance can be found by connecting a variable resistor between the terminals of the cell and measuring the p.d. across it for a range of resistances (R). The intercept of the line on the I axis of a graph of I against R will give the internal resistance of the cell ($r = \frac{\varepsilon}{I}$).

Exam practice

1 (a) Figure 5.23 shows the variation of current with voltage for a metal wire at two different temperatures.

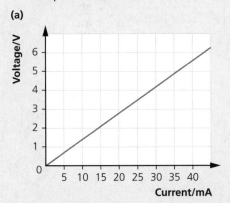
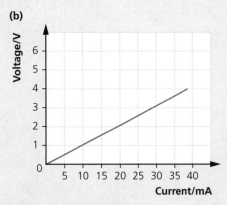

(a)

(b)

Figure 5.23 Voltage–current variation at different temperatures

    (i) Calculate the resistance of the wire at each temperature. [2]

    (ii) Which graph shows the higher temperature? [1]

(b) (i) What is the resistance of the component at the point marked A on the graph in Figure 5.24? [2]

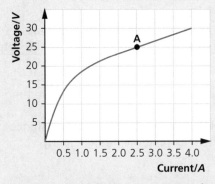

Figure 5.24 Voltage–current variation for a component

    (ii) Does the material disobey Ohm's law? Explain your answer. [2]

2 (a) What is the resistance between the points A and B of the combination of resistors shown in Figure 5.25? [2]

(b) Explain how you arrived at the answer. [4]

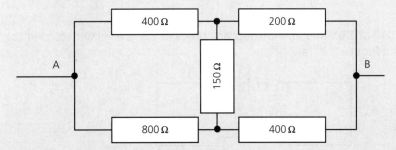

Figure 5.25 Resistance network

3 You are given four resistors, 20 kΩ, 10 kΩ, 5 kΩ and 1 kΩ. How would you connect two or more of them to make the following total resistances?

    (a) 15 kΩ [1]

    (b) 14 kΩ [2]

    (c) 6.67 kΩ [2]

    (d) 4.33 kΩ [2]

4 A current I flows though a wire of length L and radius of cross-section r, which is made of material of resistivity ρ.

The rate of heat generation in the wire is:

A $\dfrac{IL\rho}{r}$

B $\dfrac{I^2L\rho}{\pi r^2}$

C $\left[\dfrac{L\rho}{\pi r^2}\right]^2$

D $\dfrac{\pi r^2}{I\rho L}$ [1]

5 A potential divider is set up as shown in Figure 5.26.

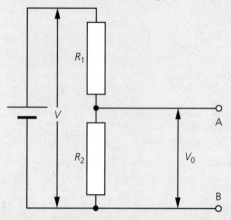

Figure 5.26 **Potential divider circuit**

The emf of the cell is 6 V and the values of R_1 and R_2 are 200 Ω and 400 Ω respectively. A digital voltmeter of very high resistance (>10 MΩ) connected between A and B is used to measure the output voltage (V_0).

(a) Calculate the output voltage. [2]

(b) If the digital voltmeter is replaced with an analogue meter of resistance 1000 Ω calculate the new output voltage. Explain your answer. [3]

(c) The digital voltmeter is replaced and R_1 is replaced by an NTC thermistor of initial resistance 200 Ω. Explain what happens to the output voltage when the thermistor is heated gently. [2]

6 (a) What is the definition of a volt? [1]

(b) What is the definition of electromotive force (emf)? [1]

(c) What is meant by internal resistance? [1]

(d) Why is internal resistance of a source a useful safety factor? [2]

7 A low-voltage school power supply has an emf of 12 V and internal resistance of 3 Ω. Calculate the currents drawn from the power supply and the values of the output voltage when the power supply is connected to:

(a) a resistor of 25 Ω [2]

(b) a resistor of 2.5 Ω. [1]

(c) Explain your answers. [2]

8 Three identical cells, each with an emf of 1.5 V and an internal resistance of 2.0 Ω, are connected in series to a 4.0 Ω resistor, as shown in Figure 5.27 (a).

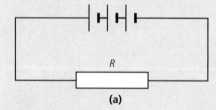

(a)

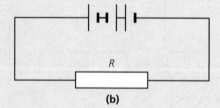

(b)

Figure 5.27

If one of the cells is reversed (Figure 5.27(b)) what is the ratio of the power output in R in circuit (a) to the power output in R in circuit (b)?

A   3.0

B   5.4

C   7.2

D   9.0   [1]

Answers and quick quiz 5 online

ONLINE

Summary

You should now have an understanding of:

- the basics of electricity — this to include charge and current as a flow of charge, resistance and Ohm's law and the variation of current with voltage for a number of circuits
- current–voltage characteristics — various versions of these curves can be found in Figure 5.4 (p. 96)
- resistivity (ρ) — this is a property of the material and not a particular specimen; resisitivity = resistance × area/length (units Ω m)

- resistance and temperature — how the resistance of resistors and thermistors changes with temperature
- circuits — combinations of resistors in series and parallel, and their associated formulae
- potential divider circuits — the use of two or more resistors to give a fractional output of the applied p.d.
- electromotive force and internal resistance — the emf (ε) of the cell is the maximum potential difference that the cell can produce across its terminals; the output potential difference is less due to the internal resistance of the cell

6 Further mechanics and thermal physics

Periodic motion

Circular motion

REVISED

Motion in a circular path

An object moves in a circle (Figure 6.1) at constant linear speed (v) and constant angular velocity (ω). The linear velocity is constantly changing because the direction of the linear motion is changing.

$$\text{average angular velocity } (\omega_{av}) = \frac{\theta}{t}$$

$$\text{instantaneous angular velocity } (\omega_{inst}) = \frac{\Delta \theta}{\Delta t}$$

The angular velocity (ω) is measured in radians per second.

$$\text{linear velocity } (v) = \text{angular velocity } (\omega) \times \text{radius of the circle } (r)$$

$$\text{period of the motion } (T) = \frac{2\pi}{\omega} = \frac{2\pi r}{v}$$

$$= \frac{1}{\text{number of revs per second } (f)}$$

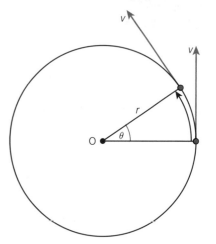

Figure 6.1 Motion in a circular path

Angular *speed* can be written in terms of the frequency (f) of the motion as:

$$\text{angular speed } (\omega) = \frac{v}{r} = 2\pi f$$

Example

Calculate the angular velocity of an object that makes:
(a) 25 revs per second
(b) 33.3 revs per minute

(Give your answers in radians per second.)

Answer

(a) $T = \frac{1}{f} = 4 \times 10^{-2}\,\text{s}$

$\omega = \frac{2\pi}{T} = \frac{6.283}{4 \times 10^{-2}} = 157\,\text{rad s}^{-1}$

(b) $T = \frac{1}{f} = \frac{60}{33.3} = 1.80$

$\omega = \frac{2\pi}{T} = \frac{6.283}{1.80} = 3.49\,\text{rad s}^{-1}$

Exam tip

In circular motion the angle of rotation is usually measured in radians rather than degrees. The radian is the angle subtended at the centre of a circle by an arc equal to the radius of the circle. Therefore $360° = 2\pi$ radians and so 1 radian (1^c) is $57.3°$.

Now test yourself

TESTED

1 An object moves at a constant speed of $5\,\text{m s}^{-1}$ in an orbit of radius $3\,\text{m}$.
   (a) What is the period of the motion?
   (b) How many revolutions per second does the object make?
   (c) What is its angular velocity?
   (d) What is its angular acceleration?

Answer on p. 219

Centripetal acceleration and centripetal force

An acceleration towards the centre of the circle occurs because the direction of the linear velocity is changing and is known as the **centripetal acceleration**.

$$\text{centripetal acceleration } (a) = \frac{v^2}{r} = \omega^2 r$$

Since the mass is accelerating, there must be a force acting on it. This force acts towards the centre of the circle and is known as the **centripetal force**. Therefore, applying Newton's second law gives:

$$\text{centripetal force } (F) = \frac{mv^2}{r} = m\omega^2 r$$

Example

The force of friction between a certain car and the road is 10 000 N. If the mass of the car is 1000 kg what is the maximum speed at which it can take a corner of radius 40 m?

Answer

$$\text{force} = 10\,000 = \frac{mv^2}{r} = \frac{1000 \times v^2}{40}$$

$$v^2 = \frac{10\,000 \times 40}{1000} = 400$$

$$v = 20\,\text{m s}^{-1}$$

Notice that it is the mass of the car that is used in the equation and *not* its weight. If the speed is greater than this maximum value, the frictional force will not be able to retain the original radius.

Now test yourself

TESTED

2 An object moves in a circle with a constant period. Which of the following correctly describes its motion?
  ● constant speed
  ● constant velocity
  ● acceleration with constant magnitude
  ● constant accelerating force
3 A stone of mass 4 kg is tied to a string and swung in a horizontal circle of radius 2 m with a speed of 4 m s⁻¹.
  (a) What is the centripetal force on the stone?
  (b) How many revolutions does the stone make every second?
  (c) In what direction will the stone move at the moment just after the string is cut?
  (d) What will the force become if the radius of the orbit is halved, with the speed of the stone remaining constant?

Answers on p. 219

Motion in a vertical circle

If a bucket of water is whirled round in a circle the bucket and water are continually accelerating towards the middle of the circle. If the circle is in a vertical plane the water does not fall out if the bucket is moving at a sufficiently high velocity, because its centripetal acceleration will be greater than *g*.

Space stations and artificial gravity

The rotation of a space station is used to create the sensation of gravity at the rim. The 'floor' of a room in a rotating space station would be the outer edge of the space station, and the rotation rate to give an acceleration equal to Earth's gravity (g) will vary depending on the size of the station:

$$\text{rotation rate } (\omega) = \sqrt{\frac{g}{r}}$$

Example

An astronaut is spun in a horizontal centrifuge with a diameter of 6 m. What must his velocity be so that his maximum acceleration is $8g$? (Take $g = 9.81 \, \text{m s}^{-2}$)

Answer

acceleration due to circular motion = $8g = 78.48 \, \text{m s}^{-2}$

centripetal acceleration = $\dfrac{v^2}{r}$

So:

$v = \sqrt{ar} = \sqrt{78.48 \times 3} = 15.3 \, \text{m s}^{-1}$

> **Exam tip**
>
> Make sure that you use the correct SI units.

> **Typical mistake**
>
> Not remembering to use the radius instead of the diameter for circular motion calculations.

Now test yourself

TESTED

4 (a) Calculate the acceleration of a satellite orbiting the Earth in a circular orbit at a distance of 42 000 km from its centre with an orbit time of 1 day (86 400 s).
  (b) What might such a satellite be used for?

Answer on p. 219

Carousel fairground rides

The angle (θ) that the wires make with the vertical and the radius of their 'orbit' (r) depends on the angular velocity (ω) alone and not the mass of the chairs.

$$\tan \theta = \frac{r\omega^2}{g}$$

Simple harmonic motion

REVISED

$$\text{acceleration } (a) = -kx$$

where k is a constant and x is the displacement of the body from the fixed central point (O) at any time t.

Note that the maximum speed occurs at the centre of the oscillation while the maximum acceleration occurs at the two 'ends' of the oscillation (maximum displacement).

The maximum displacement of the body on either side of its central position is called the amplitude (A). The period of the motion (T) is the time it takes for the body to make one complete oscillation.

The **frequency** (f) is the number of complete oscillations per second, or $1/T$.

> A body is undergoing **simple harmonic motion** (SHM — Figure 6.2) if it has an acceleration that is:
> ● directed towards a fixed point, and
> ● proportional to the displacement of the body from that point.

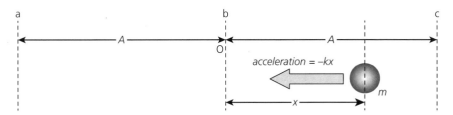

Figure 6.2 Simple harmonic motion

Simple harmonic motion equations

The equation for SHM is usually written as:

$$\text{acceleration } (a) = -\omega^2 x$$

where ω is a constant ($= 2\pi f$), and x is the displacement of the body from the fixed point (O) at any time t. The value of ω depends on the particular system of oscillation.

- **acceleration** $(a) = -\omega^2 x$
- **displacement** from fixed point $(x) = A\cos\omega t$ $x = A$ when $t = 0$
- velocity and acceleration:
  - $v = A\omega\sin\omega t$
  and
  - $a = -A\omega^2\cos\omega t$
- **velocity** $(v) = \pm\omega\sqrt{A^2 - x^2}$
- **period** $(T) = \dfrac{2\pi}{\omega}$
- maximum speed $= \omega A$ (when $x = 0$)
- maximum acceleration $= \omega^2 A$ (when $x = \pm A$)

> **Exam tip**
>
> The value of t is measured with $t = 0$ being taken at one end of the oscillation (maximum displacement) and *not* the centre.

Example

A body oscillates with simple harmonic motion with an amplitude of 12 cm and a frequency of 20 Hz. Calculate:

(a) the maximum acceleration of the mass
(b) the maximum velocity of the mass
(c) the displacement from the centre 0.02 s after leaving one end of the oscillation
(d) the velocity 0.02 s after leaving one end of the oscillation

Answer

(a) $x = A = 0.12$ m
$a = -\omega^2 x$
$\omega = 2\pi f$
So $\omega = 127$ rad s^{-1}
$a = -127^2 \times 0.12 = -1.94 \times 10^3$ m s^{-2}
(b) $v = r\omega = 0.12 \times 127 = 15.2$ m s^{-1}
(c) displacement from one end $= A\cos\omega t = 0.12\cos(127 \times 0.02) = 0.1$ m
(d) velocity $= A\omega\sin\omega t = 0.12 \times 127 \times \sin(127 \times 0.02) = 8.63$ m s^{-1}

Now test yourself

TESTED ☐

5 A body oscillates with simple harmonic motion. The displacement (x) from the centre of the oscillation at a time t (measured from when the object is in the centre) is $x = 0.03\sin(2\pi t/3)$. (All numerical values are in the appropriate SI units.)
  (a) What is the amplitude of the motion?
  (b) What is the period of the motion?
  (c) What is the frequency of the motion?
  (d) Calculate the displacement 0.5 s after the object passes the mid point of the oscillation

Answer on p. 219

Simple harmonic motion graphs

Velocity is the gradient of the displacement–time graph. Acceleration is the gradient of the velocity–time graph (Figure 6.3).

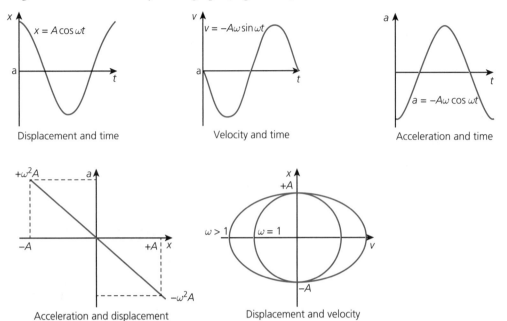

Displacement and time

Velocity and time

Acceleration and time

Acceleration and displacement

Displacement and velocity

Note: These graphs show $x = A$ when $t = 0$

Figure 6.3 Simple harmonic motion graphs

Answer on p. 219

Now test yourself

TESTED

6 At what points of the oscillation shown in Figure 6.2 is:
   (a) the displacement of the mass numerically greatest
   (b) the velocity of the mass numerically greatest
   (c) the velocity of the mass numerically least
   (d) the acceleration of the mass numerically least
   (e) the acceleration of the mass numerically greatest?

Simple harmonic systems

REVISED

The helical spring

A mass m is suspended at rest from a helical (spiral) spring (Figure 6.4) such that the extension produced is e. If the spring constant is k we have: $mg = ke$. Units for k = newtons per metre (N m^{-1}).

If the mass is then pulled down a small distance x and released, the mass will oscillate due to both the effect of the gravitational force downwards (mg) and the varying upward force from the spring ($k(e + x)$).

The resulting acceleration of the mass (a) is $-kx/m$. This shows that the acceleration is directly proportional and oppositely directed to the displacement, and so the motion is simple harmonic.

From the defining equation for simple harmonic motion ($a = -\omega^2 x$) we have $\omega^2 = k/m = g/e$ and therefore the period of the motion T for a helical spring is given by:

$$\text{period } (T) = 2\pi\sqrt{\frac{m}{k}} = 2\pi\sqrt{\frac{e}{g}}$$

where g is the gravitational acceleration.

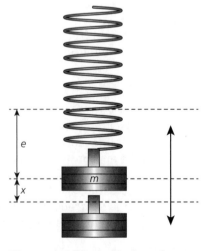

Figure 6.4 The helical spring

A mass of 0.4 kg oscillates with simple harmonic motion with an amplitude of 5 cm and a frequency of 100 Hz. Calculate:
(a) the period of motion of the mass
(b) the maximum acceleration of the mass
(c) the maximum velocity of the mass

Answer

(a) period of motion $(T) = \frac{1}{f} = \frac{1}{100} = 0.01\,s$

(b) $a = -\omega^2 A$

$\omega = 2\pi f$ so $\omega = 628.3$ radians s^{-1}

$a = -628.3^2 \times 0.05 = -1.97 \times 10^4\,m\,s^{-2}$

(c) $v = A\omega = 0.05 \times 628.3 = 31.42\,m\,s^{-1}$

> **Exam tip**
>
> The displacement when the mass is released in the above example will be the amplitude of the resulting motion.

Now test yourself

TESTED

7 An elastic spring extends by 1 cm when a small mass is attached at the lower end. If the weight is pulled down by 0.25 cm, calculate the period of the resulting motion. $(g = 9.8\,m\,s^{-2})$

8 An elastic spring has a spring constant of 2.5 N m^{-1}. Calculate the mass supported by the spring if the resulting period of motion is 2 s.

Answers on p. 219

The simple pendulum

A simple pendulum of length L with a mass m attached to the lower end is displaced through an angle θ from the vertical (Figure 6.5). The restoring force F is the component of the weight of the bob towards the equilibrium position.

$$\text{period of a simple pendulum } (T) = 2\pi\sqrt{\frac{L}{g}}$$

Note that the period of a simple pendulum does not depend on the mass of the bob. This formula is only accurate for small angles of swing.

> **Typical mistake**
>
> Not remembering that for a complete oscillation the system must return to the exactly the same conditions as those at the start of the oscillation.

> **Exam tip**
>
> A simple pendulum is one with a 'weightless' string, so that all the mass of the pendulum is concentrated in the pendulum bob.

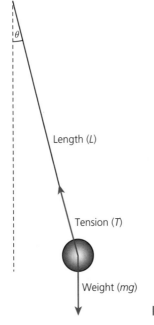

θ

Length (L)

Tension (T)

Weight (mg)

Figure 6.5 The simple pendulum

Example

What is the length of a simple pendulum that would have a period of 2 s? ($g = 9.8\,\mathrm{m\,s^{-2}}$)

Answer

period of a simple pendulum $(T) = 2\pi\sqrt{\dfrac{L}{g}}$

So:

$$L = \frac{gT^2}{4\pi^2} = 9.8 \times \frac{4}{4\pi^2} = \frac{9.8}{\pi^2} = 1.0\,\mathrm{m}$$

> **Exam tip**
>
> You should quote your answer with the correct units and give some comments about the accuracy and problems in your experiment.

Now test yourself

TESTED ☐

9 If a pendulum clock is taken to the top of the mountain does it gain or lose? Explain your answer.

10 Calculate the period of a simple pendulum of length 25 cm on the surface of Mars where $g = 3.8\,\mathrm{m\,s^{-2}}$.

Answers on p. 219

Required practical 7

Investigation into simple harmonic motion using a mass–spring system and a simple pendulum

Mass–spring system

For a range of values of masses on the spring from about 0.1 kg to 0.7 kg measure the extension of the spring (e) and the period of oscillation (T). The value for T should be found by measuring ten oscillations (10 T) and then the time for one worked out.

Check that the spring will return to its original length after each load is added, by removing that load.

For each mass calculate the value of T^2. Plot a graph of e against T^2 and hence determine g using the graph and the formula $T^2 = \dfrac{4\pi^2 e}{g}$.

A best-fit line should always be plotted — this does not necessarily include the first or last point. However, in this experiment, the line must go through (0, 0) because with no load the period would be zero.

Simple pendulum

A simple pendulum can be used to measure the acceleration due to gravity (g). A pendulum bob is tied to the end of a thread, which is suspended from a clamp. The length of the pendulum is measured and then the bob is displaced though a small distance and the time for ten oscillations (10 T) is recorded. From this the period (T) can be calculated, and hence T^2. The period is then measured for a series of different values of L (from 0.2 m to 1.2 m) and a graph plotted of T^2 against L.

The gradient of this graph is equal to $\dfrac{g}{4\pi^2}$. Therefore:

$$g = \frac{4\pi^2 L}{T^2}$$

As with the spring, a best-fit line should always be plotted — this does not necessarily include the first or last point.

Energy considerations in simple harmonic motion

The kinetic energy of any body of mass m and velocity v is $\frac{1}{2}mv^2$.

So, in simple harmonic motion:

$$\text{kinetic energy } (E_K) = \frac{1}{2}m\omega^2(A^2 - x^2)$$

Now the maximum value of the kinetic energy will occur when $x = 0$, and this will be equal to the total energy of the body. Therefore:

$$\text{total energy} = \frac{1}{2}m\omega^2A^2$$

Therefore, since potential energy = total energy − kinetic energy, the potential energy at any point will be given by:

$$\text{potential energy } (E_P) = \frac{1}{2}m\omega^2x^2$$

Graphs of the variation of potential energy, kinetic energy and the total energy are shown in Figure 6.6.

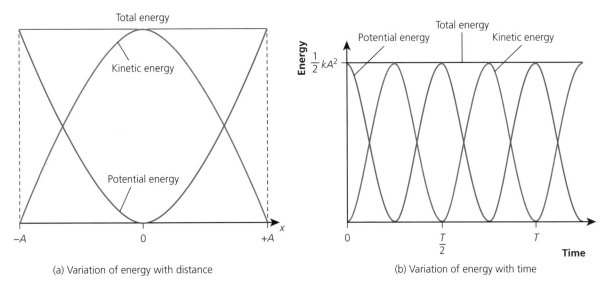

(a) Variation of energy with distance

(b) Variation of energy with time

Figure 6.6 Energy in simple harmonic motion

Example

A mass of 0.3 kg is suspended on a helical spring. When displaced through a vertical distance of 8 cm and released, it oscillates with a frequency of 10 Hz. Calculate:
(a) the period of motion of the mass
(b) the velocity of the mass 0.5 s after release from its lowest point
(c) the kinetic energy of the mass at this point

Answer
(a) Period of motion $(T) = \frac{1}{f} = \frac{1}{10} = 0.1\,\text{s}$

(b) $\omega = 2\pi f = 2\pi \times 10 = 6.28\,\text{rad s}^{-1}$
velocity $= A\omega\sin(\omega t) = 0.08 \times 6.28 \times \sin(6.28 \times 0.05)$
$= 0.08 \times 6.28 \times 0.00055 = 0.00275\,\text{m s}^{-1}$

(c) kinetic energy $= \frac{1}{2}mv^2 = \frac{1}{2} \times 0.3 \times 0.00275^2$
$= 1.13 \times 10^{-6}\,\text{J}$

Typical mistake

Not remembering to convert centimetres to metres when calculating energy in joules.

Now test yourself

TESTED

11 A trolley of mass 1.5 kg is placed on a horizontal, frictionless surface and attached to two supports by helical springs so that it can oscillate with simple harmonic motion. If the horizontal distance between the two ends of the motion is 30 cm and the period is 1.8 s:
    (a) Calculate the maximum kinetic energy of the trolley during the motion.
    (b) State at which point in the oscillation this occurs.

Answer on p. 219

> **Exam tip**
>
> Make sure that you use the correct SI units.

Forced vibrations and resonance

REVISED

Free, damped and forced oscillations

There are three main types of simple harmonic motion: free, damped and forced oscillations.

Free oscillations

In free oscillations (Figure 6.7) the amplitude remains constant as time passes — there is no damping. In other words there is no loss of energy from the oscillator to the surroundings. This type of oscillation will only occur in theory, because in practice there will always be some energy transfer.

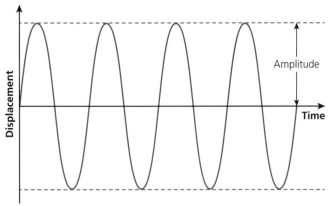

Figure 6.7 Free oscillations

Damped oscillations

Damping (Figure 6.8) means a loss of energy from the oscillator to the surroundings. This produces a decreasing amplitude and varying period due to external or internal damping forces.

Damping can occur as:
- natural damping, for example internal forces in a spring, and fluids exerting a viscous drag
- artificial damping, for example electromagnetic damping in galvanometers and top-pan balances, the coating of panels in cars to reduce vibrations, and shock absorbers in cars

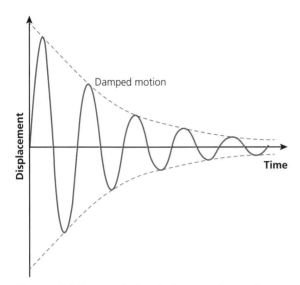

Figure 6.8 Damped simple harmonic motion

Forced oscillations

These are vibrations that are driven by an external force. A simple example is a child's swing: as you push it, the amplitude increases. A loudspeaker is also an example of forced oscillation; it is made to vibrate by the varying force between the field produced by the current in the speaker coil and that of a fixed magnet.

Resonance

Forced vibrations can also show another very important effect. With a child's swing, you will find that if you push in time with the natural frequency of the swing then the oscillations build up rapidly. This last fact is an example of **resonance**.

All systems have their own natural frequency. If you apply a driving force of the same frequency as the natural frequency and in phase with the initial oscillations, then resonance results — the amplitude of the oscillations gets larger and larger.

No mechanical system will vibrate at only its resonant frequency — harmonic oscillations will also occur. The amount of damping of a system affects the shape of the resonance curve. (Figure 6.9).

> **Exam tip**
>
> Heavily damped systems give broad resonance curves while lightly damped systems give sharply peaked resonance curves.

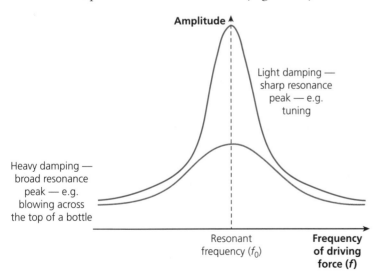

Figure 6.9 Resonance curves for different levels of damping

> **Revision activity**
>
> Make a mind map about the differences in resonance for a variety of oscillating systems.

Now test yourself

12 When pushing a swing, if the driving force has a frequency half that of the resonant frequency, will resonance result? Explain your answer.

Answer on p. 219

Exam practice

1 An astronaut is travelling round the Earth in a circular orbit where the acceleration due to gravity is g'. His acceleration towards the Earth is:
   A 2g' B g' C zero   D −g [1]

2 A heavy mass attached to a light, inextensible string moves in a horizontal circle at a constant speed. Which of the following is true? [1]
   A The mass is not subject to any accelerating force.
   B The kinetic energy of the mass is constant.
   C If the string is cut, the mass would move outwards radially.
   D The angular momentum of the mass varies sinusoidally.

3 A stone of mass 4 kg is tied to a string and swung in a horizontal circle of radius 2 m, with a speed of 4 m s^{-1}.
   (a) What is the force on the stone? [2]
   (b) How many revolutions does the stone make every second? [2]
   (c) In what direction will the stone move if the string is cut? [1]
   (d) What will the force become if the radius of the orbit is halved? [1]

4 An astronaut is spun in a vertical centrifuge with a radius of 3 m. What must his velocity be so that his maximum acceleration is 7g? (Take $g = 9.8$ m s^{-2}) [3]

5 Show that when a ball is swung in a vertical circle on a piece of string, at a constant linear velocity, the minimum velocity of the ball such that the tension in the string is zero:
   (a) occurs at the top of the circle [2]
   (b) is independent of the mass of the ball [1]
   (c) is equal to $\sqrt{rg}$, where g is the gravitational acceleration and r is the radius of the circular orbit [1]

6 Which of the following are simple harmonic motion? Explain your answers.
   (a) the vibration of a tuning fork [1]
   (b) an elastic super ball bouncing on the ground [1]
   (c) a large rectangular box, resting on the floor, that is slightly tilted and then released [1]
   (d) a trampolinist bouncing up and down on a trampoline [1]
   (e) a simple pendulum [1]
   (f) a mass fixed to a helical spring oscillating up and down [1]
   (g) a ball being swung round in a horizontal orbit on the end of a piece of string [1]

7 A particle oscillates with simple harmonic motion with a period of 2 s. If its maximum velocity is 0.4 m s^{-1} calculate:
   (a) its velocity when its displacement is half the amplitude [2]
   (b) its acceleration when its displacement is one quarter of the amplitude [2]

8 If a simple pendulum with a period of 2.00 s on the surface of the Earth ($g = 9.8$ m s^{-2}) were taken to the surface of the Moon where the gravitational acceleration is 1.62 m s^{-2}, what would be the value of its period?
   A 2.00 s   B 1.62 s   C 4.90 s   D 6.05 s   [1]

9 The period of a certain simple pendulum is 2.0 s and the mass of the pendulum bob is 50 g. The bob is pulled aside through a horizontal distance of 8 cm and then released. Find the displacement and kinetic energy of the bob 0.7 s after its release. [3]

10 Discuss why the separation of 'rumble strips' placed across a road might be important. [2]

Answers and quick quiz 6 online

Thermal physics

Thermal energy transfer

Internal energy

> The internal energy of a material is composed of **kinetic energy** and **potential energy**.
>
> **Kinetic energy** — vibrational energy in a liquid and a solid; vibrational and translational energy in a liquid and a gas.
>
> **Potential energy** due to attraction between molecules — large in a solid, smaller in a liquid and assumed to be zero in an ideal gas except during the actual collisions between gas molecules.

The graph in Figure 6.10 shows the force and potential energy variation between two molecules separated by a distance r. In a solid, the point M is their equilibrium position when they are separated by a distance r_0.

The first law of thermodynamics

The internal energy of a system is increased (ΔU) when thermal energy (ΔQ) is transferred to it by heating or when work is done by it (ΔW). This statement is known as the first law of thermodynamics:

$$\Delta U = \Delta Q - \Delta W$$

Note that ΔU represents both the change in the internal kinetic energy (an increase in molecular velocity) and the increase in the internal potential energy (due to an increase in energy overcoming intermolecular forces as a result of separation of the molecules). The potential energy increase is zero for ideal gases and negligible for most real gases, except at temperatures near liquefaction and/or at very high pressures.

If the change occurs without a change of temperature ($\Delta U = 0$) it is called an isothermal change, and so the first law becomes:

$$\Delta Q - \Delta W = 0$$

Heat capacity

The units of thermal capacity are joules per kelvin ($J\,K^{-1}$).

Specific heat capacity

$$\text{specific heat capacity } (c) = \frac{\Delta Q}{m\Delta\theta}$$

where ΔQ is the heat energy input, m is the mass of the body and $\Delta\theta$ is the rise in temperature.

The units of specific heat capacity are joules per kilogram kelvin ($J\,kg^{-1}\,K^{-1}$).

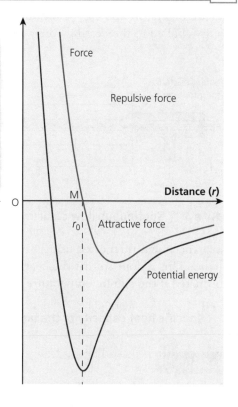

Figure 6.10 Forces between molecules

> **Exam tip**
>
> There is a finite amount of energy in the universe. This energy can be transferred from one form to another; the total amount never changes – if we want to use energy in one form then we have to 'pay for it' by converting it from energy in another form.

> The **thermal capacity** or **heat capacity** of a body is the heat energy needed to raise its temperature by 1 K.

> The **specific heat capacity** (c) of a material is the heat energy needed to raise the temperature of 1 kg of the material by 1 K.

Continuous flow calorimeter

Liquid flows in from a constant−head apparatus at a constant rate past a thermometer, where the temperature (θ_0) is recorded. It then flows around the heater coil and out past a second thermometer where the outlet temperature (θ_1) can be measured (Figure 6.11). The electrical energy applied to the heating coil is E_1 and the liquid flow rate m_1.

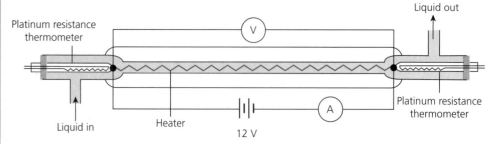

Figure 6.11 **Continuous flow calorimeter**

The experiment is repeated using a new electrical energy (E_2) and flow rate m_2. This is done so that the heat loss during the experiment can be eliminated if the rise in temperature is the same in each experiment.

$$\text{specific heat capacity of the liquid }(c) = \frac{E_2 - E_1}{(m_2 - m_1)(\theta_1 - \theta_0)}$$

Example 1

A block of metal of mass 0.5 kg initially at a temperature of 100°C is gently lowered into an insulated copper container of mass 0.05 kg containing 0.9 kg of water at 20°C. If the final temperature of the mixture is 25°C, calculate the specific heat capacity (c) of the metal of the block. (Assume no loss of heat and that no water is vaporised.) (specific heat capacities: water 4200 J kg^{-1} K^{-1}; copper 385 J kg^{-1} K^{-1})

Answer

heat lost by block = $0.5 \times c \times (100 - 25) = 37.5c$

heat gained by water and container = $(0.9 \times 4200 \times 5) + (0.05 \times 385 \times 5)$
= 18 996 J

Therefore:

$37.5c = 18\,996$ J

specific heat capacity (c) = $\dfrac{18\,996}{37.5} = 506.6$ J kg^{-1} K^{-1}

Example 2

Water flows through a continuous flow calorimeter at 150 g per minute. When the heater power is adjusted to 40 W the difference between the inlet and outlet temperatures is 3 K. When the flow rate is increased to 450 g per minute the heater power has to be increased to 100 W to maintain the same temperature difference. Calculate the specific heat capacity of water.

Answer

specific heat capacity = $\dfrac{100 - 40}{(7.5 - 2.5) \times 10^{-3} \times 3} = 4000$ J kg^{-1} K^{-1}

Typical mistake

Forgetting to convert grams to kilograms and minutes to seconds in specific heat capacity calculations.

13 A hot copper rivet of mass 150 g is dropped into 250 g of water initially at 16°C. If the water temperature rises to 35°C what was the initial temperature of the rivet? (specific heat capacities: water 4200 J kg⁻¹ K⁻¹; copper 385 J kg⁻¹ K⁻¹)

14 How does the high value of the specific heat capacity of water help to reduce the variation in temperature of land masses adjacent to oceans?

Answers on p. 219

Change of state and specific latent heat

During a change of state the potential energies of the particles involved are changing but their kinetic energy remains constant. Therefore no temperature change occurs during the change of state.

The **heat energy** (ΔQ) required to change the state of a mass (m) of a material of specific latent heat L is:

$$\text{quantity of heat energy } (\Delta Q) = mL$$

Figure 6.12a shows how the temperature of a specimen might alter with time due to a steady heat input — heat losses to the exterior have been ignored here. Figure 6.12b shows how the molecular arrangements within the material change as the heat energy is supplied.

> The **specific latent heat** (L) is the energy required to change the state of 1 kg of a substance.

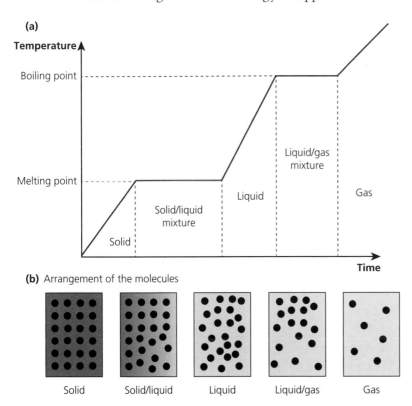

Figure 6.12 Change of state

Every material has two specific latent heats: **specific latent heat of fusion** and **specific latent heat of vaporisation**.

> The **specific latent heat of fusion** is the heat energy needed to change 1 kg of the material in its solid state at its melting point to 1 kg of the material in its liquid state, and that released when 1 kg of the liquid changes to 1 kg of solid.
>
> The **specific latent heat of vaporisation** of a liquid is the heat energy needed to change 1 kg of the material in its liquid state at its boiling point to 1 kg of the material in its gaseous state, and that released when 1 kg of vapour changes to 1 kg of liquid.

Example 1

How much heat energy is needed to heat 250 g of water at 15°C to steam at 100°C? (specific heat capacity of water, 4200 J kg⁻¹ K⁻¹; specific latent heat of vaporisation of water = 2.25×10^6 J kg⁻¹)

Answer

heat energy input $= mc\Delta\theta + mL$

$= (0.25 \times 4200 \times 85) + (0.25 \times 2.25 \times 10^6) = 89\,250 + 562\,500 = 651\,750$ J

Example 2

Calculate the amount of ice that would be melted by a 65 W heater in 5 minutes at 0°C if all other heat energy exchanges are ignored.

Answer

specific latent heat of fusion of ice $= 330\,000$ J kg⁻¹

electrical energy input $= 65 \times 5 \times 60 = 19\,500$ J

mass of ice melted $= 19\,500/330\,000 = 0.059$ kg $= 59$ g

Now test yourself

TESTED

15 How long will it take for a 60 W electrical heater immersed in an ice–water mixture at 0°C to raise the temperature of the mixture to 20°C if all other heat energy exchanges are ignored? The mixture initially contains 200 g of water and 25 g of ice. (specific latent heat of fusion of ice = 330 000 J kg⁻¹; specific heat capacity of water = 4200 J kg⁻¹ K⁻¹)

16 Why is a scald due to steam at 100°C much more dangerous than one due to boiling water at 100°C?

Answer on p. 220

Ideal gases

REVISED

> An **ideal gas** is one where the molecules are considered to be infinitely small and do not exert any force on each other. It also obeys Boyle's law and has an internal energy that is dependent only on the temperature of the gas.

The gas laws

Boyle's law

The pressure of a fixed mass of gas is inversely proportional to its volume as long as its temperature remains constant:

$$p_1V_1 = p_2V_2$$

Exam practice answers and quick quizzes at **www.hoddereducation.co.uk/myrevisionnotes**

Pressure law

The pressure of a fixed mass of gas is directly proportional to its absolute temperature (in kelvin) as long as its volume remains constant:

$$\frac{p_1}{T_1} = \frac{p_2}{T_2}$$

Charles's law

The volume of a fixed mass of gas is directly proportional to its absolute temperature (in kelvin) as long as its volume remains constant:

$$\frac{V_1}{T_1} = \frac{V_2}{T_2}$$

These gas laws are represented in Figure 6.13.

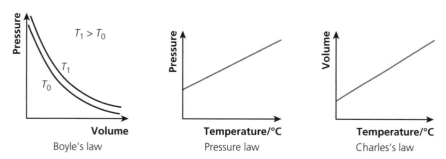

Figure 6.13 Gas law graphs

Example 1

A fixed mass of gas at a constant temperature and a pressure of 10^5 Pa is compressed from 2 litres to 500 cm³. Calculate the new pressure of the gas.

Answer

Using Boyle's law, $p_1V_1 = p_2V_2$:

$$10^5 \times 2 \times 10^{-3} = p_2 \times 500 \times 10^{-6}$$

Therefore:

$$p_2 = \frac{10^5 \times 2 \times 10^{-3}}{500 \times 10^{-6}} = 4 \times 10^5 \, \text{Pa}$$

A decrease in volume by a factor of four produces an increase in pressure by a factor of four.

Example 2

During a spacewalk an astronaut moves from the shadow of the spacecraft into full sun. The temperature of his oxygen tank rises from 200 K to 350 K. If the original pressure of the gas was 2×10^5 Pa, what is its new pressure?

Answer

Using the pressure law:

$$p_2 = \frac{2 \times 10^5 \times 350}{200} = 3.5 \times 10^5 \, \text{Pa}$$

Example 3

A fixed mass of gas at a constant pressure and with an initial volume of 5 litres is heated from 20°C to 80°C. Calculate the new volume of the gas.

Answer

Using Charles's law, $\dfrac{V_1}{T_1} = \dfrac{V_2}{T_2}$:

$$\frac{5 \times 10^{-3}}{293} = \frac{V_2}{353}$$

$$V_2 = \frac{353 \times 5 \times 10^{-3}}{293} = 6 \times 10^{-3} \, m^3 = 6 \text{ litres}$$

Now test yourself

TESTED

17 (a) A gas is stored in a cylinder at a pressure of 8 atmospheres (8 times the pressure of the atmosphere). If the volume of the cylinder is 3000 cm³, what is the volume of the gas that comes out of the cylinder when the pressure in the cylinder is reduced to 1 atmosphere?

   (b) The air in a closed cylinder at a pressure of 50 000 Pa and at 27°C is heated to 227°C. What is its new pressure?

18 If 4 litres of gas initially at 20°C is heated to 100°C at a constant pressure, what will be the new volume of the gas? Give your answer in m³.

Answers on p. 220

The absolute zero of temperature

The centre diagram in Figure 6.13 shows the variation of the pressure of an ideal gas with its temperature. If we draw the line back to where it cuts the temperature axis we reach a point where the pressure of the gas is zero — in other words, the molecules have stopped moving. They have no velocity and so no kinetic energy. This is the lowest temperature that it is possible to reach and is called **absolute zero**, defined as $0\,K$, or $-273.15°C$ (Figure 6.14). In fact the third law of thermodynamics states that it is impossible to actually reach this temperature. When an object is cooled, its internal energy is reduced. As the temperature approaches absolute zero it becomes more and more difficult to lower the temperature further. We are always left with what is known as 'zero-point energy', and so we can define absolute zero as the temperature at which substances have a minimum internal energy.

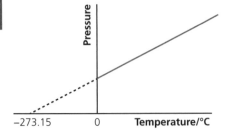
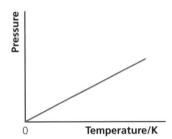

Figure 6.14 **The concept of absolute zero**

Exam practice answers and quick quizzes at **www.hoddereducation.co.uk/myrevisionnotes**

Now test yourself

19 A sample of gas is found to have a volume of 64.0 cm³ at 0°C and 87.5 cm³ at 100°C. Using these results, calculate a value for absolute zero in °C.

Answer on p. 219

The equation of state for an ideal gas

The **equation of state for an ideal gas** is:

$$\frac{pV}{T} = \text{constant}$$

For 1 mole of gas the constant is the molar gas constant, R ($R = 8.31\,\text{J}\,\text{kg}^{-1}\,\text{mol}^{-1}$).

The **ideal gas equation** (for n moles) is:

$$pV = nRT$$

> **Exam tip**
>
> Standard temperature and pressure (STP) is sometimes referred to in questions. Standard temperature is taken as 0°C (273.15 K) and standard pressure as $1.01 \times 10^5\,\text{Pa}$.

Now test yourself

20 Calculate the number of moles of a gas of pressure 10^5 Pa, at a temperature of 27°C, and occupying a volume of 5 m³.

Answer on p. 220

Gas constants and their relationships

The **Avogadro constant** (N_A) = 6.02×10^{23} particles per mole

The **molar gas constant** (R) = $8.31\,\text{J}\,\text{mol}^{-1}\,\text{K}^{-1}$

The **Boltzmann constant** (k) = $1.38 \times 10^{-23}\,\text{J}\,\text{K}^{-1}$

For an ideal gas of n moles and N molecules:

number of molecules = **number of moles** × **Avogadro constant** $\quad N = nN_A$

Boltzmann constant = $\dfrac{\text{molar gas constant}}{\text{Avogadro constant}} \quad k = \dfrac{R}{N_A}$

In 1 mole of hydrogen (2 g) there are 6.02×10^{23} molecules (hydrogen exists as H_2).

In 1 mole of oxygen (32 g) there are 6.02×10^{23} molecules (oxygen exists as O_2).

In 1 mole of copper (63 g) there are 6.02×10^{23} atoms.

In 1 mole of uranium-235 (235 g) there are 6.02×10^{23} atoms.

For example, if we have 2 kg of uranium in a fuel rod we have $\dfrac{2000}{235}$ = 8.51 moles, and this contains $8.51 \times 6.02 \times 10^{23} = 5.12 \times 10^{24}$ atoms and so 5.12×10^{24} uranium nuclei.

Molar mass and molecular mass

It is important to distinguish between molar and molecular masses. The molar mass (M) is the mass of one mole while the molecular mass (m) is the mass of one molecule. For m kg of a gas of molar mass M kg, r is a further constant that depends on the gas under consideration:

$$pV = \frac{mRT}{M} = mrT$$

The Boltzmann constant

$$pV = NkT$$

where N is the number of molecules in the gas and k is the Boltzmann constant ($k = 1.38 \times 10^{-23}\,\text{J K}^{-1}$).

Work considerations during gas expansion and compression

An ideal gas at a pressure p is enclosed in a cylinder of cross-sectional area A. A force F is then applied to the piston, pushing the piston in a distance Δx and compressing the gas, decreasing its volume by a small amount, ΔV (Figure 6.15). This force is applied slowly and steadily so that, at all times, $FA = pV$ (i.e. equilibrium).

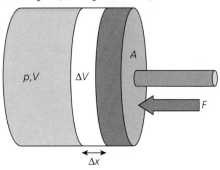

Figure 6.15 **Work done on a gas**

The force on the piston $F = pA$

So:

work done on the gas during compression $= \Delta W = pA\Delta x = p\Delta V$

work done on the gas $= p\Delta V$

If the gas expands by an amount ΔV the work done *by* the gas is also $p\Delta V$.

> **Example**
>
> An ideal gas at a pressure of 2×10^5 Pa is enclosed in a cylinder with a volume of 6 litres by a frictionless piston. How much work is done by the gas if it expands slowly and steadily to 6.05 litres?
>
> Answer
> work done by the gas $= p\Delta V = 2 \times 10^5 \times (6.05 - 6.00) \times 10^{-3} = 10$ J

Typical mistake

Forgetting to convert volumes in other units (such as cm³ and litres) to cubic metres.

Now test yourself

21 An ideal gas has a molar mass of 40 g and a density of 1.2 kg m^{-3} at 80°C. What is its pressure at that temperature?

22 Calculate the number of molecules in 5 litres of air at STP (273.15 K, 1.01 × 10^5 Pa). The Boltzmann constant = 1.38 × 10^{-23} J K^{-1}

Answers on p. 220

Required practical 8

Investigation of Boyle's and Charles's laws

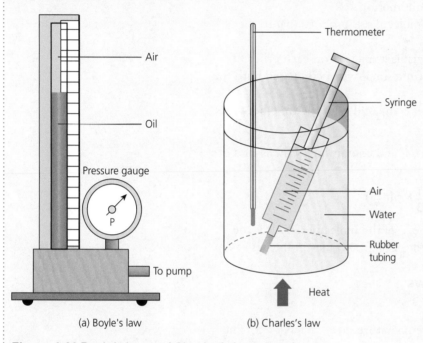

(a) Boyle's law   (b) Charles's law

Figure 6.16 Boyle's law and Charles's law experiments

Boyle's law can be verified using the standard Boyle's law apparatus shown in Figure 6.16a. Use a pump to increase the pressure of the trapped air to a maximum safe value. Place a safety screen between yourself and the apparatus. Take readings of the pressure (p) from the gauge and the volume (V) using the scale beside the tube. Slowly release the pressure in steps, recording the pressure and the volume. Allow time for any excess oil to flow down the side of the tube before taking each pair of readings. Plot graphs of p against V and p against $1/V$.

Charles's law can be verified using the apparatus shown in Figure 6.16b. The syringe is placed in water in a beaker and the beaker is then slowly heated. The volume of the air in the syringe is measured at a series of temperatures from 0°C to 100°C.

Plot graphs of V against T using the range 0°C to 100°C with a scale marked in kelvin and extrapolate back to a point where $V = 0$.

Molecular kinetic theory model

REVISED

Brownian motion

Observing a weak solution of milk, and later pollen grains in suspension, with a high-powered microscope the Scottish physicist Robert Brown saw that the particles of milk and the pollen grains showed a violent and random motion.

A simple modern version of Brown's experiment is the smoke cell. A small cell of air is placed under a microscope and illuminated strongly from the side. Some smoke is then blown into it. Through the microscope the particles of smoke can be seen to be in violent random motion (Figure 6.17). This motion is due to the collisions of the (invisible) air molecules with the much larger particles of smoke. Heating the cell makes the smoke particles' motion even more violent, due to the increased velocity of the air molecules.

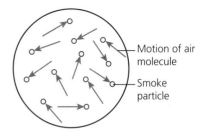

Motion of air molecule

Smoke particle

Figure 6.17 Brownian motion

Kinetic theory model for an ideal gas

In deriving the formula for the kinetic theory of gases, the following assumptions are made about molecules:

- molecules behave as if they were hard, smooth, elastic spheres
- molecules are in continuous random motion
- the average kinetic energy of the molecules is proportional to the absolute temperature of the gas
- molecules do not exert any appreciable attraction on each other
- the volume of the molecules is infinitesimal when compared with the volume of the gas
- the time spent in collisions is small compared with the time between collisions

For an ideal gas at pressure p, volume V and containing N molecules of mass m:

$$pV = \frac{1}{3} Nm(c_{rms})$$

where c_{rms} is the root mean square speed of the molecules. This is the average of all the squares of the speeds of the molecules in the gas.

Kinetic theory and the gas laws

Boyle's law

For a given mass of gas at constant temperature, $pV = \frac{1}{3} mN(c_{rms})^2$. The total mass of gas is $mN = M$, and therefore:

$$pV = \frac{1}{3} M(c_{rms})^2$$

which is constant, and this is Boyle's law.

Charles's law

The ideal gas equation for n moles of gas is $pv = nRT$ and so for 1 mol of gas we have $pv = RT$, where R is the gas constant. But in 1 mol of gas there are N_A molecules, where N_A is the Avogadro constant, and therefore:

$$pV = \frac{1}{3} mN_A c^2 = RT$$

But if the temperature of the gas is changed from T_1 to T_2 with a resulting change in volume from V_1 to V_2, the pressure being kept constant:

$$pV_1 = \frac{1}{3} M(c_{1rms})^2 \text{ and } pV_2 = \frac{1}{3} M(c_{2rms})^2$$

Therefore:

$$\frac{V_1}{V_2} = \frac{(c_{1rms})^2}{(c_{2rms})^2}$$

But kinetic energy $= \frac{1}{2}mc_{rms}^2$ and therefore:

$$\frac{V_1}{V_2} = \frac{\text{kinetic energy}_1}{\text{kinetic energy}_2}$$

One of the assumptions of kinetic theory is that the kinetic energy of the molecules is directly proportional to the absolute temperature of the gas (T) and therefore:

$$\frac{V_1}{V_2} = \frac{(c_{1rms})^2}{(c_{2rms})^2} = \frac{T_1}{T_2}$$

So the volume is directly proportional to the absolute temperature, and this is Charles's law.

Example

The density of nitrogen at STP = 1.251 kg m^{-3}. Calculate the root-mean-square speed (c) of nitrogen molecules.

Answer

pressure = 1.013×10^5 Pa

$$c^2 = \frac{3p}{\rho} = \frac{3 \times 1.013 \times 10^5}{1.251} = 2.432 \times 10^5$$

$$c = 493 \, \text{m s}^{-1}$$

Now test yourself

23 Calculate the root-mean-square speeds for the following gases at a pressure of 10^5 Pa:
   (a) air, density 1.29 kg m^{-3}
   (b) carbon dioxide, density 1.98 kg m^{-3}
   (c) nitrogen, density 1.25 kg m^{-3}
   (d) chlorine, density 3.21 kg m^{-3}
   (e) hydrogen, density 0.09 kg m^{-3}
24 A sample of gas of volume 0.1 m^3 and at a pressure of 2.0×10^5 Pa is enclosed in a cylinder. If the root-mean-square speed of the gas molecules is 5.5×10^2 m s^{-1} and the mass of each molecule is 3.5×10^{-26} kg calculate the number of gas molecules in the cylinder.

Answers on p. 220

Kinetic energy of a molecule

The average kinetic energy of the randomly moving molecules, each of mass m, in a gas is directly proportional to the absolute temperature of the gas (T).

$$\frac{1}{2}mc^2 = \frac{3}{2}kT = \frac{3RT}{2N_A}$$

where k is the Boltzmann constant (= 1.38×10^{-23} J K^{-1}).

Example

Calculate:

(a) the kinetic energy of an individual gas molecule of mass 3.5×10^{-26} kg moving at a speed of $600\,\mathrm{m\,s^{-1}}$

(b) the average kinetic energy of the gas molecules in a cylinder at a temperature of 20°C

Boltzmann constant = $1.38 \times 10^{-23}\,\mathrm{J\,K^{-1}}$

Answer

(a) $\frac{1}{2}mv^2 = \frac{1}{2} \times 3.5 \times 10^{-26} \times 600^2 = 6.3 \times 10^{-21}$ J

(b) $\frac{3}{2}kT = \frac{3}{2} \times 1.38 \times 10^{-23} \times 293 = 6.07 \times 10^{-21}$ J

Now test yourself

 TESTED

25 Calculate the average kinetic energy of a molecule in a gas at a temperature of 300 K. Ideal gas constant = $8.3\,\mathrm{J\,mol^{-1}\,K^{-1}}$; Avogadro number = $6.04 \times 10^{23}\,\mathrm{mol^{-1}}$.

Answer on p. 220

Exam practice

1 (a) A copper saucepan of mass 250 g is filled with 850 g of water at 18°C. A 500 W heater is then placed in the water and switched on for 5 minutes. What is the theoretical temperature of the water after this time? (Specific heat capacities: water $4200\,\mathrm{J\,kg^{-1}\,K^{-1}}$; copper $3570\,\mathrm{J\,kg^{-1}\,K^{-1}}$) [3]

  (b) Why will the actual rise be less than this? [1]

2 In a determination of the specific heat capacity of water using a continuous flow calorimeter the following readings were taken:

First experiment:
- electrical power supplied to the heater = 20 W
- 0.431 kg of water flowed through the calorimeter in 20 minutes
- rise in temperature of the water = 11.5°C

Second experiment:
- electrical power supplied to the heater = 23.8 W
- 0.524 kg of water flowed through the calorimeter in 20 minutes
- rise in temperature of the water = 11.5°C

Calculate the specific heat capacity of water. [4]

3 A block of metal of mass m requires a heater with a power P to just keep it molten. When the heater is switched off the mass solidifies completely in a time t. If the rate of loss of heat is constant the specific latent heat of the block is:

A $\dfrac{P}{mt}$ B $\dfrac{t}{Pm}$ C $\dfrac{Pt}{m}$ D $\dfrac{Pm}{t}$ [1]

4 An insulated beaker contains 0.025 kg of ice and 0.3 kg of water at 0°C. Steam at 100°C is passed into the container until all the ice has melted. Assume that no heat is lost from the system.

  (a) How much thermal energy will the condensed steam have lost? [3]

  (b) How much thermal energy will the ice have absorbed? [1]

  (c) What will be the final mass of water in the beaker? [2]

  (Specific latent heat of ice = $3.4 \times 10^5\,\mathrm{J\,kg^{-1}}$; specific latent heat of steam = $2.3 \times 10^6\,\mathrm{J\,kg^{-1}}$; specific heat capacity of water = $4.2 \times 10^3\,\mathrm{J\,kg^{-1}\,K^{-1}}$)

5 (a) Give a short account of Brownian motion and state what molecular information can be obtained from observation of the effect in a smoke cell. [2]

  (b) Calculate the root-mean-square speed of molecules of oxygen gas at a temperature of 27°C. (Boltzmann constant = $1.38 \times 10^{-23}\,\mathrm{J\,K^{-1}}$; mass of an oxygen molecule = 5.3×10^{-26} kg) [2]

Exam practice

6 Equal masses of two gases are contained in two separate cylinders at equal pressures and temperatures. The volume of the cylinder containing gas 1, which has a molecular weight of 40 g, is 3 litres. If the volume of the cylinder containing gas 2 is 4 litres, what is the molecular weight of gas 2? [3]

7 One mole of an ideal gas at a pressure of 1×10^5 Pa and a temperature of 300 K has a volume of 0.25 m³. At 600 K and 5×10^5 Pa the volume occupied in m³ is:

A 5   B 0.1   C 50   D 10   [1]

8 The pressure p of an ideal gas is given by the formula $p = \frac{1}{3}\rho c^2$, where:

A the appropriate unit for c^2 is m²s⁻¹

B ⅓ is an approximation for $\frac{1}{\pi}$

C ρ is the mass per unit volume of the gas

D $\sqrt{c^2}$ is the average speed of the molecules [1]

Answers and quick quiz 6 online

ONLINE

Summary

You should now have an understanding of:

- Circular motion — angular speed, centripetal force
- Simple harmonic motion (SHM) — conditions for SHM, SHM equations, graphical representation
- Simple harmonic systems — mass–spring system, simple pendulum
- Forced vibrations and resonance — free, forced and damped oscillations and resonance

- Thermal energy transfer — specific heat capacity, change of state, specific latent heat
- Ideal gases — the gas laws (Boyle's, pressure and Charles's), absolute zero, work done = $p\Delta V$
- Avogadro, Boltzmann and molar gas constants
- Molecular kinetic theory — relation between p, V and T in terms of the simple molecular model, kinetic theory equation and molecular kinetic energy

Fields

Force fields

There are similarities and differences between gravitational and electric fields:
- The field strength is dependent on *both* the masses or charges.
- The field strength is proportional to the inverse square of their separation.
- Gravitational fields are *always* attractive.
- Electric fields can be attractive or repulsive.
- Electric fields are affected by the intervening medium.

> The region around a mass in which an object is affected by that mass is called a **gravitational field.**
>
> The region around an electric charge in which an object is affected by that charge is called an **electric field.**

Gravitational fields

Newton's law of gravitation

Gravitation is a universal attractive force that acts between any mass and any other mass in the universe. In 1666 Newton, proposed his universal law of gravitation. He considered a planet (mass m) moving in a circular orbit (radius r) at angular velocity ω round the Sun (mass M):

$$\text{force on a planet} = F = m\omega^2 r = m\left(\frac{2\pi}{T}\right)^2 r = \frac{4\pi^2 mr}{T^2}$$

Newton assumed an inverse square law of force between the bodies, that is:

$$F = \frac{kMm}{r^2}$$

where k is a constant. Therefore: $\dfrac{T^2}{r^3} = \dfrac{4Tr^2}{k}$

which is also constant and so $T^2 \propto r^3$, which agrees with Kepler's third law of planetry motion.

The gravitational force between two bodies, of mass m and M and with their centres a distance r apart, is given by Newton's law of gravitation (Figure 7.1). This force is always attractive.

Figure 7.1 Newton's law of gravitation

$$\text{gravitational force} = \frac{GMm}{r^2}$$

where G is the universal gravitational constant and has a value of $6.67 \times 10^{-11}\,\text{N}\,\text{m}^2\,\text{kg}^{-2}$.

Example

Calculate the force between the following:
(a) a mass of 100 kg and one of 200 kg placed 2.5 m apart
(b) the Earth and the Sun

(Mass of the Earth = 6×10^{24} kg; mass of the Sun = 2×10^{30} kg; mean distance r of the Earth from the Sun = 1.5×10^{11} m)

Answer

Using $F = GMm/r^2$ and $G = 6.67 \times 10^{-11}$ N m^2 kg^{-2} gives:
(a) $F = 2.13 \times 10^{-7}$ N
(b) $F = 3.56 \times 10^{22}$ N

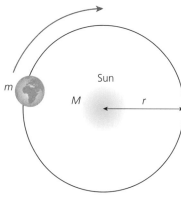

Figure 7.2 Earth in orbit round the Sun (not to scale)

It is gravitation that holds the planets in orbit around the Sun. Although the law strictly deals only with point masses, the sizes of the planets are so small compared with their distances of separation that we can consider them to be point masses. (This also applies to the Sun itself in this context.)

Now test yourself

TESTED

1 What is the force between two bodies of mass:
(a) 80 kg and 60 kg placed 5 m apart
(b) 2000 g and 800 kg placed 25 m apart
(c) 1.5×10^{12} kg and 3×10^{27} kg placed 45×10^9 m apart?

Answer on p. 220

Example

The Earth (mass m) orbits the Sun (mass M) in 365 days in an orbit of radius 1.5×10^{11} m. Calculate the mass of the Sun. (Remember that $v = \frac{2\pi r}{T}$ (page 114) and use $G = 6.67 \times 10^{-11}$ N m^2 kg^{-2}.)

Answer

$$\frac{mv^2}{r} = \frac{m(4\pi^2 r^2/T^2)}{r} = \frac{GMm}{r^2}$$

Therefore:

$$\frac{4\pi^2}{T^2} = \frac{GM}{r^3}$$

$$M = \frac{4\pi^2 r^3}{GT^2} = 4\pi^2 \times \frac{(1.5 \times 10^{11})^3}{6.67 \times 10^{-11} \times (86\,400 \times 365)^2}$$

$$= \frac{1.33 \times 10^{35}}{6.63 \times 10^4} = 2.0 \times 10^{30}\,\text{kg}$$

Exam tip

Make sure that you use the correct units for gravitation problems.

Gravitational field strength

REVISED

The **gravitational field strength** (g) is the force on a unit mass at a point in the field.

The force (F) on a body of mass m in a gravitational field of a body of mass M is $F = \dfrac{GMm}{r^2}$. Force per unit mass is given by $\dfrac{F}{m}$. So:

$$\text{gravitational field strength } (g) = \frac{F}{m} = \frac{GM}{r^2}$$

The units for g are $N\,kg^{-1}$ or, since $F = ma$, $\dfrac{F}{m} = a$ and so g can be expressed in $m\,s^{-2}$.

The gravitational field strength at the surface of a body of radius R is written as g_o. Therefore, the gravitational field strength (g) at any other point distance r from the centre is $g = \dfrac{g_o R^2}{r^2}$.

Uniform and radial gravitational fields

In a uniform gravitational field the field lines are parallel and of equal separation.

In a radial gravitational field the field lines get further apart the further from the central mass (Figure 7.3).

Uniform field — parallel field lines   Radial field — converging field lines

Figure 7.3 Uniform and radial gravitational fields

Example

Use the following data to calculate the gravitational field strength at the surface of the Earth. (Assume the Earth to be a uniform sphere.)

Mass of Earth = 6.0×10^{24} kg; radius of Earth = 6.4×10^6 m; gravitational constant (G) = $6.67 \times 10^{-11}\,N\,m^2\,kg^{-2}$

Answer

gravitational field strength $(g) = \dfrac{GM}{R^2} = \dfrac{6.67 \times 10^{-11} \times 6.0 \times 10^{24}}{(6.4 \times 10^6)^2} = 9.8\,N\,kg^{-1}$

Gravitational potential

REVISED

Near the Earth's surface the gravitational potential changes uniformly with distance. The gravitational potential in a radial field varies with the distance from the central mass. At a distance r from a central mass M:

$$\text{gravitational potential} = -\frac{GM}{r}$$

The gravitational potential energy of a mass m at a distance r from the centre of the central body is:

$$\text{gravitational potential energy} = -\frac{GMm}{r}$$

Note the negative sign in both the above equations. This is because the zero of gravitational potential for a mass is taken to be at an infinite distance from that mass.

Gravitational field strength (g) at a point in a field is the negative gravitational potential gradient at that point:

$$g = -\frac{\Delta V}{\Delta r}$$

Gravitational field lines and equipotentials can be represented diagrammatically (Figure 7.4).

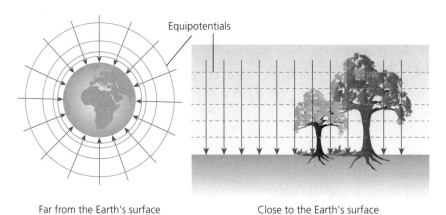

Far from the Earth's surface   Close to the Earth's surface

Figure 7.4 Gravitational field lines and equipotential surfaces

Change in the gravitational potential energy of an object

If an object of mass m moves from point A (a distance r_1 from the centre of a radial field due to a mass M) to a point B (a distance r_2 from the centre of the field), its potential energy will change. The change in the gravitational potential energy of an object (ΔE_p) is given by:

$$\Delta E_p = GMm\left(\frac{1}{r_1} - \frac{1}{r_2}\right)$$

This change is independent of the path that the mass takes between A and B.

Now test yourself

TESTED

2 A satellite of mass m moves in a circular orbit of radius r about the centre of a planet of mass M, remaining at a distance h above the surface of the planet. What is the gravitational potential energy of the satellite?

A $-mgh$ B $-\frac{GMm}{r}$ C $\frac{GMm}{r}$ D mgh

Answer on p. 220

The **gravitational potential** (V) at a point in a field is defined as the work done in moving a unit mass to that point in the field.

Exam tip

The gravitational energy of a mass at the surface of a planet is negative.

Exam tip

For more on field lines and equipotentials see pages 146-147 on electric fields. The same concepts apply to gravitational fields.

Exam tip

Gravitational field lines are always at right angles to the equipotential surfaces.

Escape velocity

If we consider a space probe of mass m at the surface of a planet radius R then its gravitational potential energy at the surface is $-GMm/R$.

The energy required to escape from the field is therefore $+GMm/R$.

$$\text{escape velocity } (v_e) = \sqrt{\left(\frac{2GM}{R}\right)} = \sqrt{(2Rg_0)}$$

The escape velocity is also important if we want to find whether a planet can retain its atmosphere. The higher-velocity molecules will escape the gravitational pull if their velocity is greater than the escape velocity of the planet.

> The **escape velocity** of a planet, or any other gravitational system, is the velocity that an object would have to be given to escape from the gravitational field of that planet.

Now test yourself

TESTED

3 Explain why the maximum velocity with which a meteoroid could enter the Earth's atmosphere is equal to the escape velocity of the Earth, assuming no forces act other than the gravitational attraction between the meteoroid and the Earth.

Answer on p. 220

Orbits of planets and satellites

REVISED

Geostationary and geosynchronous satellites

A **geosynchronous** satellite has a period of exactly 1 day so that it passes over a fixed point on the Earth's surface at the same time each day but does not remain constantly over that point.

A special case is the **geostationary** satellite (Figure 7.5), which gets its name from the fact that it is launched into an **equatorial orbit** with a period of exactly 1 day and so when viewed from any point on the Earth's surface it remains constantly in a fixed position. This type of satellite is used for communications.

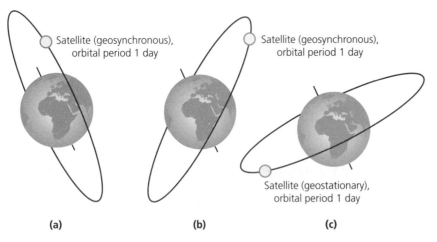

Figure 7.5 Geosynchronous (a–c) and geostationary (c) satellite orbits

The radius (r) of the orbit for a geostationary satellite is:

$$r = \frac{g_0 R^2 T^2}{4\pi^2}^{1/3}$$

where g_o is the surface gravity for the planet, R its radius and T the orbital period of the satellite.

Geostationary satellites (for the Earth) have the following properties:
- Their orbit lies in the same plane as the equator of the Earth.
- The satellite period is one Earth day.
- They have the same angular velocity as the Earth.
- They move around their orbit in the same direction as the rotation of the Earth.
- They remain above a fixed point on the Earth's surface.
- They have an orbit centred on the centre of the Earth.

> **Typical mistake**
>
> Using the orbit height and not the orbit radius when making calculations concerning satellite orbits.

Now test yourself

TESTED ☐

4 Figure 7.6 shows four orbits that have been suggested for a communications satellite. Which orbit(s) is(are) both possible and correct for this use? Explain your answer.

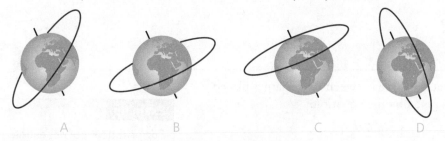

Figure 7.6 **Possible geostationary satellite orbits**

5 What is the speed of a satellite placed in orbit around the Earth 200 km above the surface of the Earth? (radius of the Earth = 6.4×10^6 m, mass of the Earth = 6×10^{24} kg, $G = 6.67 \times 10^{-11}$ N m^2 kg^{-2})

Answers on p. 220

Exam practice

1 A planet moves in a circular orbit of radius r about the centre of a star. Its period of rotation about the star is proportional to:

    A $r^{\frac{1}{2}}$ **B** $r^{\frac{3}{2}}$ **C** $r^{-\frac{1}{2}}$ **D** $r^{-\frac{3}{2}}$ [1]

2 Calculate the mass of the Earth using only the following data: $G = 6.67 \times 10^{-11}$ N m^2 kg^{-2}, radius of the Moon's orbit = 3.8×10^8 m, orbit time of the Moon about the Earth = 27.3 Earth days. [3]

3 A 1600 kg satellite is launched from the surface of a planet that has a uniform density, a surface gravity of 30 N kg^{-1} and a radius of 2×10^7 m. It is placed into a circular orbit 1.5×10^6 m above the planet's surface such that the satellite has a period of 5×10^3 s.

    (a) Give an estimate of the approximate increase in potential energy of the satellite. [2]
    (b) Explain why the actual answer will differ from this. [2]
    (c) Calculate the exact answer. [2]
    (d) Calculate the kinetic energy of the satellite in this orbit. [3]

4 A satellite is launched so that is in a geostationary orbit. Calculate the height of the satellite above the equator using only the following data: value of g at the Earth's surface = 9.81 m s^{-2}, equatorial radius of the Earth = 6.4×10^6 m. [2]

5 Triton, one of the Moon's of Neptune, has a mass of 2.14×10^{22} kg and a diameter of 2700 km. Calculate:

    (a) the surface gravity of Triton [2]
    (b) the escape velocity of Triton [3]

Answers and quick quiz 7 online

ONLINE ☐

Summary

You should now have an understanding of:
- Fields — bodies in a field experience a non-contact force
- Vector representation of force fields
- Similarities and differences between gravitational and electric fields
- Gravity as a universal attractive force between all matter
- Gravitational fields — Newton's law of gravitation:

$$F = \frac{GMm}{r^2}$$

- Gravitational field strength — g as a force per unit mass:

$$g = \frac{GM}{r^2}$$

- Gravitational potential — zero at infinity
- Work done by moving a mass in a field — independent of the path of the movement
- Gravitational potential in a radial field:

$$V = -\frac{GM}{r}$$

- Orbits of planets and satellites — geostationary and geosynchronous satellites and their orbits

Electric fields

Coulomb's law

REVISED ☐

Coulomb's law states that that the force between two charges placed a distance r apart in a vacuum is given by the equation:

$$F = \left(\frac{1}{4\pi\varepsilon_0}\right)\left(\frac{Q_1Q_2}{r^2}\right)$$

The constant ε_0 is known as the **permittivity of free space**. Its value is $8.85 \times 10^{-12}\,\mathrm{F\,m^{-1}}$. The numerical value of $1/(4\pi\varepsilon_0)$ is 8.99×10^9.

> **Exam tip**
>
> For practical purposes the permittivity of air is virtually the same as that of free space (a vacuum).

> **Example**
>
> Calculate the force between the proton and electron in an atom of hydrogen. (radius of the electron orbit = $10^{-10}\,\mathrm{m}$; charge on the electron = $1.6 \times 10^{-19}\,\mathrm{C}$)
>
> **Answer**
>
> $$F = \left(\frac{1}{4\pi\varepsilon_0}\right)\left(\frac{Q_1Q_2}{r^2}\right) = \frac{8.99 \times 10^9 \times 1.6 \times 10^{-19} \times 1.6 \times 10^{-19}}{10^{-20}} = 2.3 \times 10^{-8}\,\mathrm{N}$$

Electric field strength

REVISED ☐

Electric field lines in uniform and radial electric fields

An electric field may be represented by a diagram that shows field lines (Figure 7.7). These are lines that show both the direction in which a positive charge would move in the field and also the strength of the field. The closer the lines the stronger (the more intense) the electric field at that point.

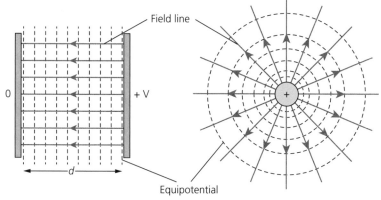

Figure 7.7 **Uniform and radial electric fields**

In a uniform field the electric field lines are equally spaced while in a radial field they get further apart at points further from the charge. Note that the dotted lines are lines of equal potential.

$$\text{electric field strength} = \frac{F}{Q}$$

force (F) on a charge (Q) in an electric field of strength $E = EQ$

The units for electric field strength are N C^{-1}, or V m^{-1}.

In a uniform field:

$$\text{electric field strength } (E) = \frac{V}{d}$$

In a radial field:

$$\text{electric field strength due to a charge } Q \ (E) = \left(\frac{1}{4\pi\varepsilon_0}\right)\left(\frac{Q}{r^2}\right)$$

The **electric field strength** (E) at a point in an electric field is the force (F) on a unit positive charge placed at that point.

Exam tip

Unlike magnetic fields an electric field will exert a force on stationary charged particles within the field.

Trajectory of a charge in an electric field

If a moving charged particle enters a uniform electric field at right angles the trajectory followed will be parabolic (Figure 7.8).

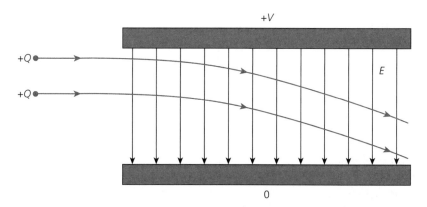

Figure 7.8 **Trajectory of a charged particle in an electric field**

Just like an object moving in a gravitational field the charged particle accelerates in the direction of the field but moves at a constant speed at right angles to it.

Now test yourself

6 Calculate the acceleration of an electron moving in a field of intensity $1000\,V\,m^{-1}$ and state in which direction this acceleration takes place. (charge on the electron = $1.6 \times 10^{-19}\,C$; electron mass = $9 \times 10^{-31}\,kg$)

Answer on p. 220

Electric potential

Note: an alternative definition is that it is the potential energy of a unit positive charge placed at that point with the zero being at infinity.

The electric potential (V) at a point a distance r from a charge Q is given by the equation:

$$V = \left(\frac{1}{4\pi\varepsilon_0}\right)\left(\frac{Q}{r}\right)$$

> The **electric potential** at a point in an electric field is defined as the work done in bringing a unit positive charge from infinity to that point (units: volts).

If a charge q is placed in an electric field, its electric potential energy is qV:

Electric potential energy at a point in a radial field $= qV = \left(\frac{1}{4\pi\varepsilon_0}\right)\frac{qQ}{r}$

Now test yourself

7 Calculate the electric potential at a distance of $0.5\,m$ from a charge of $0.5\,mC$.

Answer on p. 220

The electric field strength can also be defined as the negative of the electric potential gradient:

$$\text{electric field strength } (E) = -\frac{\Delta V}{\Delta r}$$

Hence the units $V\,m^{-1}$.

The change in electric potential is equal to the area between two distances on the E–r graph. (See Figure 7.9.)

$$\Delta V = E\Delta r$$

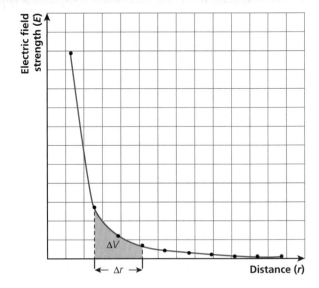

Figure 7.9 Potential, potential gradient and electric field

Equipotential lines and surfaces

A line of equipotential is a line joining points of equal potential just like a contour line on a map joining points of equal height. Since $E = -\Delta V/\Delta r$, the closer the lines of equipotential are the greater the potential gradient and so the greater the field strength.

An equipotential surface is a surface that joins points of equal potential, and therefore a surface on which the potential does not vary. This means that if you connected a wire between two points on the surface no current would flow between them.

Exam tip

Electric field lines are always at right angles to lines of equal potential and to equipotential surfaces.

No work is done in moving an electric charge along an equipotential surface.

Figure 7.10 shows equipotential lines for uniform and radial fields.

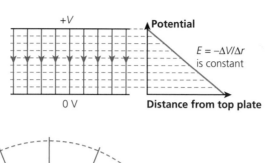

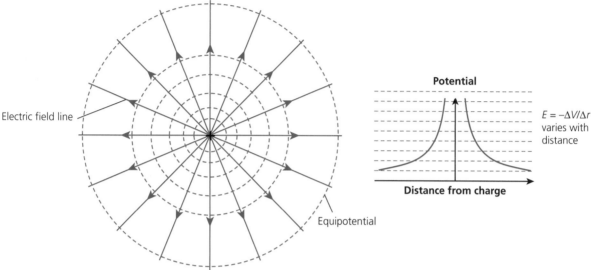

Figure 7.10 Equipotential lines for uniform and radial fields

Capacitance

Capacitance

The combination of any two conductors separated by an insulator is called a **capacitor**. A capacitor is a device that can be made to store electric charge.

Practical capacitors come in three basic forms:
- as part of an integrated circuit
- as two parallel plates (described more fully later)
- as a cylinder made of a pair of rolled–up plates

The ability to store charge is called the **capacitance** of the capacitor.

$$\text{capacitance } (C) = \frac{\text{charge } (Q) \text{ on one of the conductors}}{\text{potential difference } (V) \text{ between the two conductors}}$$

Capacitance is measured in farads (F).

A farad is actually a very large unit. The capacitors that you will meet will have capacitances of microfarads (μF, 10^{-6} F), nanofarads (nF, 10^{-9} F) or picofarads (pF, 10^{-12} F).

> A **capacitor** has a capacitance of 1 farad if the potential across it rises by 1 volt when a charge of 1 coulomb is stored in it.

Example 1

A 4700 µF capacitor is connected to a cell so that the potential difference across its terminals is 6 V. When it is fully charged:
(a) What is the charge on the positive plate of the capacitor?
(b) How many additional electrons are on the negative plate?

(charge on one electron = 1.6×10^{-19} C)

Answer

(a) $Q = CV = 4700 \times 10^{-6} \times 6 = 0.028$ C

(b) $\dfrac{0.028}{1.6 \times 10^{-19}} = 1.76 \times 10^{17}$

Example 2

What is the capacitance of a capacitor that stores 30 µC of charge when a potential difference of 6 V is applied across it?

Answer

$$\text{capacitance } (C) = \frac{\text{charge } (Q)}{\text{pd } (V)}$$

$$= \frac{30 \times 10^{-6}}{6} = 5 \times 10^{-6} \text{ F} = 5\,\mu\text{F}$$

Revision activity

Make a list of the uses capacitors both as individual components and as part of an integrated circuit.

Now test yourself

TESTED

8 (a) A capacitor of capacitance 5 µF is connected to a 6 V supply. What charge is stored in the capacitor?
 (b) A 400 pF capacitor carries a charge of 2.5×10^{-8} C. What is the potential difference across the plates of the capacitor?

Answer on p. 220

The parallel-plate capacitor

REVISED

In its most basic form a capacitor is simply two metal plates with a vacuum between them (Figure 7.11). Such an arrangement is called a parallel-plate capacitor. The two plates of a charged parallel capacitor each carry charges of the same size but of opposite sign.

$$\text{electric field strength between the plates } (E) = \frac{Q}{\varepsilon_0 A}$$

where ε_0 is the permittivity of free space (page 144).

But if V is the potential difference between the plates:

$$E = \frac{V}{d} = \frac{Q}{\varepsilon_0 A}$$

And therefore, since $C = \dfrac{Q}{V}$:

$$\text{capacitance } (C) \text{ of a parallel-plate capacitor} = \frac{\varepsilon_0 A}{d}$$

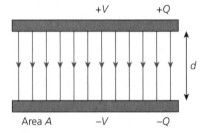

Figure 7.11 The parallel-plate capacitor

Exam tip

The electric field within a parallel-plate capacitor is the same at all points between the plates.

TESTED ☐

Now test yourself

9 Calculate the capacitance of an air-filled, parallel-plate capacitor formed of two plates, each of area 25 cm² and separated by 0.2 mm $\varepsilon_0 = 8.85 \times 10^{-12}$ F m⁻¹.

10 A school decides to make an air-filled capacitor from two metal plates separated by 0.5 mm. If the capacitance required is 100 µF what must be the area of each of the plates?

Answers on pp. 220–1

The action of a dielectric

A dielectric material is one which is non-conducting. If such a material is placed in an electric field the charges within its molecules separate, although this separation does not cause conduction (Figure 7.12).

If a dielectric material is placed between the plates of a charged capacitor, opposite charges will be induced on the two surfaces of the dielectric. This has the effect of reducing the potential difference across the capacitor. The capacitance of the capacitor is increased.

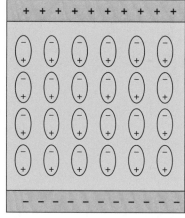

Figure 7.12 Action of a dielectric

Relative permittivity

It is often useful to compare the electrostatic properties of materials by using a quantity known as the relative permittivity (ε_r) — also known as the dielectric constant. This is defined using the relation:

permittivity (ε) = relative permittivity (ε_r) × permittivity of free space (ε_0)

The capacitance of a parallel-plate capacitor with a material of relative permittivity ε_r filling the space between the plates is:

$$\text{capacitance } (C) = \frac{\varepsilon_r \varepsilon_0 A}{d}$$

The relative permittivity of most gases is very nearly equal to 1.00, but for solids and liquids it varies between 80 (water) and 2 (paraffin).

The relative permittivity (ε_r) can also be defined as the ratio of the capacitance when the space between the plates is a dielectric to that when there is a vacuum between the plates.

> **Exam tip**
>
> Relative permittivity has no units — it is a pure number.

Example 1

Calculate the separation of the plates of an air-filled capacitor of capacitance 20 nF if the plates have an area of 45 cm². Permittivity of free space (ε_0) = 8.85 × 10⁻¹² F m⁻¹. (Take the relative permittivity of air to be 1.0)

Answer

$$\text{capacitance } (C) = \frac{\varepsilon_0 A}{d}$$

$$20 \times 10^{-9} = \frac{8.85 \times 10^{-12} \times 45 \times 10^{-4}}{d}$$

$$d = 2 \times 10^{-6}\,\text{m} = 0.002\,\text{mm}$$

> **Typical mistake**
>
> Forgetting to use SI units for plate separation and area when calculating the capacitance of a parallel-plate capacitor.

Calculate the capacitance of a parallel-plate capacitor with plates of area 50 cm² when the space between the plates is filled by a sheet of glass 0.5 mm thick.

Permittivity of free space $(\varepsilon_0) = 8.85 \times 10^{-12}\,F\,m^{-1}$; the relative permittivity of glass = 7.0

Answer

$$\text{capacitance } (C) = \frac{\varepsilon_0\varepsilon_r A}{d} = \frac{8.85 \times 10^{-12} \times 7.0 \times 50 \times 10^{-4}}{5 \times 10^{-4}}$$

$$= 6.2 \times 10^{-10}\,F = 0.62\,nF$$

Exam tip

Remember that when substituting for the area in the parallel-plate capacitor formula it is the area of *one* of the plates and *not* the total area of both.

Now test yourself

TESTED

11 Calculate the capacitance of a pair of parallel plates of area 25 cm² if they are separated by a piece of Perspex 0.1 mm thick. Take the permittivity of free space to be $8.85 \times 10^{-12}\,F\,m^{-1}$ and the relative permittivity of Perspex to be 3.5.

Answer on p. 221

Energy stored by a capacitor

REVISED

If a small charge ΔQ is taken from one plate to the other, the work done will be $V\Delta Q$ where V is the potential difference between the plates. If the initial charge on the positive plate is 0 and the final charge is Q, then the total energy gained in completely discharging the capacitor is:

$$\text{energy stored in a capacitor} = \tfrac{1}{2}\frac{Q^2}{C} = \tfrac{1}{2}CV^2 = \tfrac{1}{2}QV$$

The energy stored as a capacitor is charged is shown in Figure 7.13.

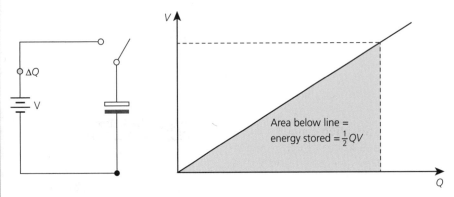

Figure 7.13 Energy stored in a capacitor

What is the energy stored in the following capacitors:
(a) a capacitor of 4700 μF charged to a potential of 240 V?
(b) a capacitor with a charge of 20 nC charged to a potential of 10 V?
(c) a capacitor of 20 μF with a charge of 50 μC?

Answer

(a) energy $= \tfrac{1}{2}CV^2 = 0.5 \times 4700 \times 10^{-6} \times 240^2 = 135.4\,J$
(b) energy $= \tfrac{1}{2}QV = 0.5 \times 20 \times 10^{-9} \times 10 = 1 \times 10^{-7}\,J = 0.1\,\mu J$
(c) energy $= \tfrac{1}{2}\frac{Q^2}{C} = 0.5 \times \frac{(50 \times 10^{-6})^2}{20 \times 10^{-6}} = 6.25 \times 10^{-5}\,J = 0.0625\,mJ$

12 A capacitor with a potential difference of 100 V between its terminals stores a charge of 5 C. What is the energy stored by the capacitor?

13 What is the energy stored in the following parallel-plate capacitor in a vacuum: area of the plates 1 m², separation 1 mm, potential difference 20 V? (permittivity of a vacuum = 8.85×10^{-12} F m⁻¹)

14 A parallel-plate capacitor is charged so that the potential difference between the plates is V. The supply is then disconnected and the plates' separation is reduced. What happens to:
  (a) the electric field strength
  (b) the charge on the plates
  (c) the potential difference between the plates?

Answers on p. 221

Capacitor charge and discharge

REVISED

When a pd is applied across the capacitor the potential cannot rise to the applied value instantaneously. As the charge on the terminals builds up to its final value it tends to repel the addition of further charge.

The rate at which a capacitor can be charged or discharged depends on:
- the capacitance of the capacitor, and
- the resistance of the circuit through which it is being charged or is discharging. The greater the resistance, the longer the time taken.

Charging a capacitor

In the charging circuit of Figure 7.14, as soon as the switch is closed in position 1 the battery is connected across the capacitor, current flows and the potential difference across the capacitor begins to rise. As more and more charge builds up on the capacitor plates, the current and the rate of rise of potential difference both fall. Finally, no further current will flow when the pd across the capacitor equals that of the supply voltage V_0. The capacitor is then fully charged.

For a capacitor charging:

$$V = V_0(1 - e^{-t/RC})$$

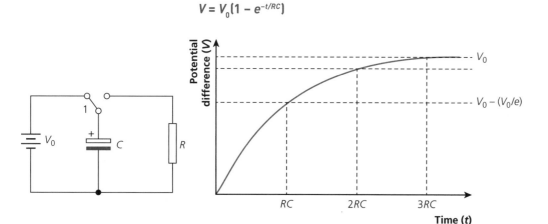

Figure 7.14 **Charging a capacitor**

A 200 µF capacitor is charged through a resistor of 100 kΩ using a 6 V DC supply. What is the potential difference across the capacitor 10 s after the commencement of the charging process?

Answer

$V = V_0(1 - e^{-t/RC}) = 6(1 - e^{-10/(100\,000 \times 0.0002)}) = 6(1 - 0.61) = 2.36\,V$

Now test yourself

TESTED

15 A 1000 µF capacitor is charged through a 2 kΩ resistor using a 12 V DC supply. How long will it take for the potential difference across the capacitor to reach 9 V?

Answer on p. 221

Discharging a capacitor

In the circuit of Figure 7.15, as soon as the switch is put in position 2 a 'large' current starts to flow and the potential difference across the capacitor drops. As charge flows from one plate to the other through the resistor the charge is neutralised and so the current falls and the rate of decrease of potential difference also falls.

Eventually the charge on the plates is zero and the current and potential difference are also zero — the capacitor is fully discharged. The value of the resistor does not affect the final potential difference across the capacitor.

$$V = V_0 e^{-t/RC} \qquad I = I_0 e^{-t/RC} \qquad Q = Q_0 e^{-t/RC} \qquad \ln V = \ln V_0 - t/RC$$

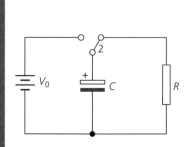

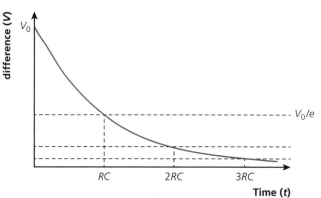

Figure 7.15 Discharging a capacitor

- The area under a current–time capacitor discharge graph gives the charge held by the capacitor.
- The gradient of a charge–time capacitor discharge graph gives the current flowing from the capacitor at that moment.
- A graph of $\ln[V_0/V]$ against t will give a linear graph with gradient $1/RC$.

Exam tip

The value of a resistor connected in series with a capacitor will not affect the final potential difference across the capacitor — only the time taken to charge or discharge it.

> **Example**
>
> A capacitor of 1000 µF with a potential difference of 12 V across it is discharged through a 500 Ω resistor. Calculate the voltage across the capacitor after 1.5 s.
>
> Answer
>
> $$V = V_0 e^{-t/RC} = 12 e^{-1.5/(500 \times 0.001)} = 0.6\,V$$

The time constant

The rate at which a capacitor charges or discharges is governed by its capacitance (C) and the resistance (R) of the circuit through which it is charged or discharged. The product RC is known as the **time constant** for the circuit:

time constant for a capacitor circuit = RC

The time constant is measured in seconds. The bigger the value of RC, the slower the rate at which the capacitor charges and discharges.

The time for the potential difference across the capacitor to halve is $T_{\frac{1}{2}} = 0.69\,RC$.

The time constant can be measured by using a graph of the discharge of a capacitor (Figure 7.15). The time taken for the potential difference to halve is found from the graph and the time constant (RC) calculated from $T_{\frac{1}{2}}/0.69$.

> **Example**
>
> The potential difference across a certain capacitor halves in 20 s Calculate:
> (a) the time constant of the capacitor
> (b) the capacitance of the capacitor if the discharge takes place through a 100 kΩ resistor
>
> Answer
>
> $$\text{time constant} = RC = \frac{\text{time for p.d. to halve}}{0.69} = \frac{20}{0.69} = 29\,s$$
>
> $$\text{capacitance } (C) = \frac{\text{time constant}}{R} = \frac{29}{100 \times 10^3} = 2.9 \times 10^{-4}\,F = 290\,\mu F$$

Now test yourself

TESTED

16 A capacitor of 1000 µF with a potential difference of 12 V across it is discharged through a 1000 Ω resistor. Calculate the voltage across the capacitor after 0.5 s.

17 What is the time constant for circuit where a 10 µF capacitor is charged through a 1000 kΩ resistor?

Answers on p. 221

Required practical 9

Investigation of the charge and discharge of a capacitor

The circuit shown in Figure 7.16 is set up with the two-way switch in the central position. The switch is moved to position 1 (Figure 7.16a) so that the capacitor charges through R_1. The potential difference across the capacitor is measured at time intervals using the high-resistance voltmeter (V) or a voltage sensor.

When the voltage is constant the switch is moved to position 2 (Figure 7.16b) so that the capacitor discharges through R_2. Once again the variation of potential difference with time is measured.

A graph of pd against time or ln (pd) against time is drawn and hence the time constant can be calculated.

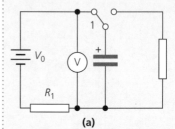 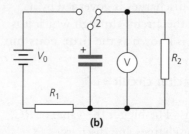

(a) (b)

Figure 7.16 Circuit diagrams for (a) charging and (b) discharging a capacitor

Exam practice

1 A parallel-plate capacitor has plates of area $6 \times 10^{-2} \, m^2$ separated by 1.2 mm. The space between the plates is filled with dielectric of relative permittivity 2.78. ($\varepsilon_0 = 8.85 \times 10^{-12} \, F \, m^{-1}$) The capacitance of the capacitor is:
   A 15.7 pF   B 2.78 nF   C 1.23×10^{-12} F   D 1.23 nF   [1]
2 A parallel-plate air capacitor of area $25 \, cm^2$ and with plates 1 mm apart is charged to a potential of 100 V.
   (a) Calculate the energy stored in it.   [2]
   (b) The plates of the capacitor are now moved a further 1 mm apart with the power supply connected. Calculate the energy change.   [3]
   (c) If the power supply had been disconnected before the plates had been moved apart, what would have been the energy change in this case?   [2]
3 To have the same energy as a 2.0 V 40 ampere-hour accumulator, a 0.1 F capacitor must be charged to a potential of:
   A 1200 V   B 2400 V   C 3600 V   D 6000 V   [1]
4 An uncharged capacitor is connected as shown in Figure 7.16a.
   (a) Describe the flow of electrons as the capacitor charges.   [1]
   (b) Describe the flow of electric current as the capacitor charges.   [1]
   (c) What determines the final potential difference across the capacitor?   [1]
   (d) The resistor R_1 is now replaced by one of larger resistance. What effect does this have on:
      (i) the rate of charging   [1]
      (ii) the final charge on the capacitor?   [1]
5 A capacitor is discharged through a 10 MΩ resistor and it is found that the time constant is 2000 s.
   (a) Calculate the value of the capacitor.   [2]
   (b) Calculate the time for the potential across a 100 µF capacitor to fall to 75 per cent of its original value if it is discharged through a 10 kΩ resistor.   [2]

Answers and quick quiz 7 online

ONLINE

Magnetic fields

Magnetic flux density

Force on a current-carrying wire

The force F on a wire carrying a current at right angles to a magnetic field (Figure 7.17a) is proportional to:

- the current in the wire I
- the length of the conductor in the field L
- the strength of the field — this is measured by a quantity known as the **magnetic flux density** (B) of the field.

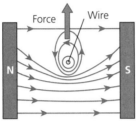

(a) View along axis of the wire

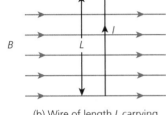

(b) Wire of length L carrying current I in a field of flux density B

Figure 7.17 Force on a wire carrying a current in a magnetic field

> Magnetic flux can be thought of as the 'flow of magnetic energy'. The amount of magnetic flux passing through a given area is the magnetic **flux density** (B) and this is measured in **teslas** (T).

With B in teslas, the force is given by the equation:

force on a wire carrying a current in magnetic field, $F = BIL$

> The **flux density** of a field of 1 tesla is the force per unit length on a wire carrying a current of one ampere at right angles to the field.

Example

Calculate the upward force on a power cable of length 150 m running east–west and carrying a current of 200 A at a place where the horizontal component of the Earth's magnetic field is 10^{-5} T.

Answer

The wire will experience a force given by
$F = BIL = 10^{-5} \times 150 \times 200 = 0.30$ N.

Now test yourself

18 A wire of length 10 cm and mass 3 g is placed at right angles to a horizontal magnetic field of strength 1.2 T. What current must be passed though the wire keep the wire at rest? ($g = 9.81$ m s^{-2})

Answer on p. 221

Fleming's left-hand rule

Professor J. A. Fleming found a simple way of remembering the direction of the force or motion using your left hand. It involves placing the thumb, first finger and second finger of the left hand at right angles to each other (Figure 7.18).

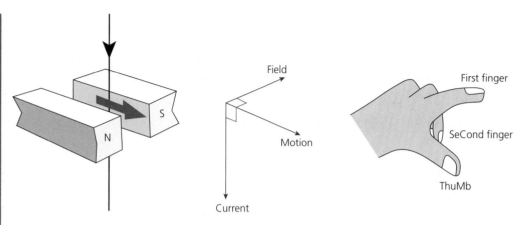

Figure 7.18 Fleming's left-hand rule

Required practical 10

Force on a current using a top pan balance

The apparatus is set up as shown in Figure 7.19 and the reading on the top pan balance is adjusted to zero. The switch is closed and a measured current passed through the rod. The force on the rod is shown on the balance. By knowing the length of rod in the field the strength (flux density) of the field can be found.

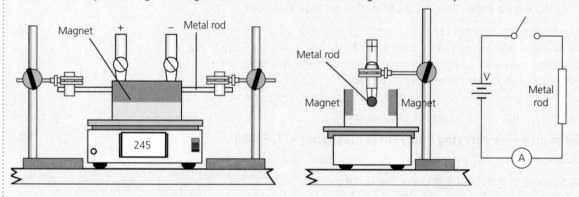

Figure 7.19 Force on a current using a top pan balance

Moving charges in a magnetic field

REVISED

Force on a charged particle moving in a magnetic field

If a wire carrying a current I is placed in a field of flux density B:

$I = nAvq$

The force (F) on the wire is given by:

$$F = BIL$$

where n is the number of charges per metre cubed, A is the cross-sectional area of the specimen, v is the drift velocity of the charges of charge q.

If a single particle moves perpendicular to the field ($\theta = 90°$ and $\sin \theta = 1$) the force on a charge q is (Figure 7.20):

$$F = Bqv$$

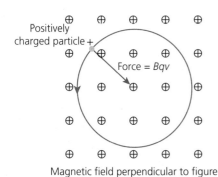
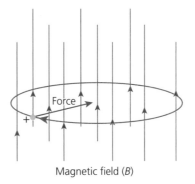

Magnetic field perpendicular to figure

Magnetic field (*B*)

Figure 7.20 Moving positive charge in a magnetic field

Since the force on a charged particle moving in a magnetic field is always at right angles to the motion (Fleming's left-hand rule) the path described will be a circle. Therefore the force on a charged particle moving a speed v perpendicular to a magnetic field of strength B is:

$$F = Bqv = \frac{mv^2}{r}$$

where r is the radius of the circular path (see centripetal force, page 115).

> **Exam tip**
>
> The force on a charged particle that is at rest in a magnetic field is zero.

> **Example**
>
> A proton (mass 1.66×10^{-27} kg) with a charge-to-mass ratio (q/m) of 9.6×10^7 C kg^{-1} moves in a circle in a magnetic field of flux density 1.2 T at 4.5×10^7 m s^{-1}.
>
> Calculate the radius of the circular orbit (see centripetal force, page 115).
>
> Answer
>
> radius of orbit $= \dfrac{mv^2}{Bqv} = \dfrac{mv}{Bq} = \dfrac{4.5 \times 10^7}{1.2 \times 9.6 \times 10^7} = 0.39$ m

Now test yourself

TESTED ☐

19 A bubble chamber photograph shows a particle moving in a circle of radius 4 mm. If the field in the chamber is 0.03 T and the velocity of the particle is 2.1×10^7 m s^{-1} what is the charge-to-mass ratio of the particle?

Answer on p. 221

Direction of force on particles of opposite sign

The force on particles of opposite sign moving in a magnetic field is in opposite directions. This is clearly shown by the motion of sub-atomic particles in a cloud chamber or bubble chamber.

The cyclotron

The force on a charged particle is fundamental to the operation of a cyclotron. This is basically a circular, evacuated chamber cut into two D-shaped halves with a high voltage across the gap. The particles are injected at the centre and a magnetic field is applied across the whole of the apparatus perpendicular to the Ds so that the particles are deflected into a circular path (Figure 7.21).

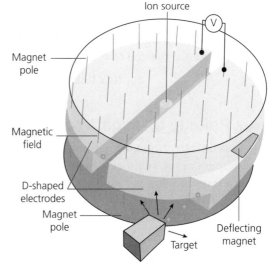

Figure 7.21 The cyclotron

They start at the centre and are given a 'kick' every time they cross the gap between the Ds, so increasing their energy.

Now test yourself
TESTED

20 Describe and explain what happens to the radius of the orbit of a charged particle as it is accelerated in a cyclotron?

Answer on p. 221

Magnetic flux and flux linkage
REVISED

Magnetic flux

Around the magnet there is a magnetic field. The magnetic flux is a measure of the magnetic field, taking into account the flux density and the extent of the field (Figure 7.22). The magnetic flux can be different at different points in the field (e.g. X and Y). Magnetic flux is given the symbol Φ and is measured in units called **webers** (Wb). The amount of flux passing through a unit area at right angles to the magnetic field lines is the flux density (B) at that point.

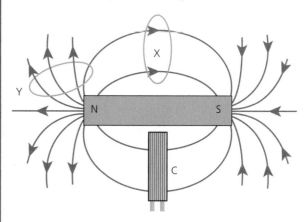

Figure 7.22 Magnetic flux and flux linkage

flux (Φ) = flux density (B) × area through which flux passes (A)

$$\Phi = BA$$

> **One weber** is the magnetic flux that, linking a circuit of one turn, produces in the circuit an emf of 1 V when the flux is reduced to zero in 1 second.

If a coil is placed at position C (Figure 7.22) the flux through the N turns is N times that through the single loop. The quantity $N\Phi$ is called the **flux linkage** for the coil at that point.

flux linkage = $N\Phi$ = NBA

Flux linkage through a rotating coil

The flux linkage through a coil that is rotating in a magnetic field changes as the angle of the coil with the magnetic field changes (Figure 7.23).

If the coil makes an angle of θ with the field normal to the plane of the coil the flux linkage is:

flux linkage = $BAN\cos\theta$

Coil of area A

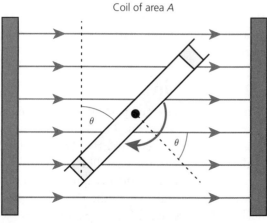

Figure 7.23 Flux through a rotating coil

Note that the axis of the coil is a line at right angles to the face of the coil.

Example

A coil of 200 turns and area 24 cm² is rotating in a magnetic field of 0.8 T. Calculate the flux linkage through the coil when the plane of the coil is at 50° to the field.

Answer

angle of axis to the field = 90 − 50 = 40°

flux linkage = $BAN \cos \theta = 0.8 \times 24 \times 10^{-4} \times 200 \times \cos 40 = 0.29$ Wb

Exam tip

Make sure that you use SI units in these calculations

Required practical 11

Flux linkage investigation — search coil and oscilloscope

The apparatus should be set up as shown in Figure 7.24.

Place the search coil in the centre of the helical spring with the plane of the coil perpendicular to the axis of the spring. Switch on the low-voltage AC supply and turn off the time base of the oscilloscope. This will give a vertical line on the screen. Determine the value of the induced emf from the length of the oscilloscope trace. Rotate the coil so that its plane is at an angle θ to the spring's axis and record the new value of the emf. Repeat this for a series of angles between 0° and 90°.

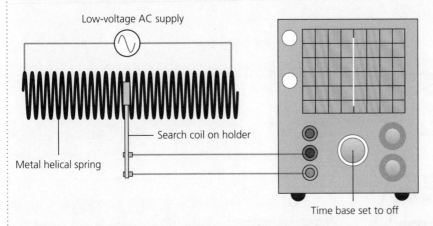

Low-voltage AC supply

Search coil on holder

Metal helical spring

Time base set to off

Figure 7.24 Flux density measurement — search coil and oscilloscope

Note that complete understanding of this experiment requires a knowledge of electromagnetic induction, which is covered in the next section.

Now test yourself

TESTED ☐

21 A coil of 100 turns and area 50 cm² is placed in a magnetic field of 0.5 T. Calculate the angle between the axis of the coil and the magnetic field if the flux linkage through the coil is 0.15 Wb.

22 A single-turn square coil with sides 20 cm long is placed with its plane at right angles to a magnetic field of strength 0.05 T.
  (a) What is the flux linkage through the coil?
  (b) The coil is now rotated about an axis through its centre, in the plane of the coil and at right angles to the field until the plane of the coil is at 30° to the field. Calculate the change of flux through the coil.

Answers on p. 221

7 Fields and their consequences

Electromagnetic induction

Applications and effects of electromagnetic induction include:

- generators
- transformers
- electromagnetic separators
- eddy current damping
- induction coils
- induction loops at approached to traffic lights

Faraday's law

If the magnetic flux through a conducting coil is altered then an emf will be generated in the coil (Figure 7.25). Faraday discovered that an emf could be generated by either:

- moving the coil or the source of flux relative to each other or
- changing the magnitude of the source of magnetic flux

Note that the emf is only produced while the flux linkage is changing.

Faraday summarised the results of his experiments as follows:

- An emf is induced in a coil if the magnetic flux through the coil changes.
- The magnitude of the induced emf depends on:
  ○ the rate of change of flux ($\Delta\Phi/\Delta t$)
  ○ the number of turns on the coil (N)

magnitude of induced emf ε = rate of change of flux linkage $= \dfrac{N\Delta\Phi}{\Delta t}$

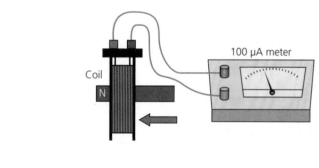

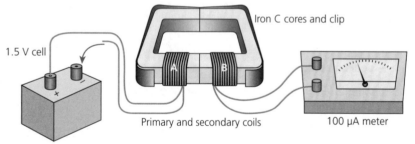

Figure 7.25 Induced emf

The faster the flux is changed the greater is the emf produced.

> ### Example
>
> The magnetic flux through a coil of 250 turns is reduced from 2.5 Wb to 0.5 Wb in 3 s. Calculate the emf produced between the two ends of the coil.
>
> #### Answer
>
> magnitude of the induced emf = rate of change of flux linkage $= \dfrac{N\Delta\Phi}{\Delta t}$
>
> $$= \frac{250 \times 2}{3} = 167\,\text{V}$$

Exam practice answers and quick quizzes at **www.hoddereducation.co.uk/myrevisionnotes**

Now test yourself

TESTED

23 It is required to produce an emf of 50 V between the two ends of a coil of 200 turns by reducing the magnetic flux through it by 2.3 Wb. Calculate the time within which this should be done.

Answer on p. 221

Fleming's right-hand rule

Fleming proposed a simple rule for giving the direction of the induced emf and current in one of Faraday's experiments (Figure 7.26).

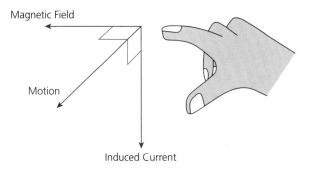

Magnetic Field

Motion

Induced Current

Figure 7.26 Fleming's right-hand rule

If the thumb and first two fingers of the right hand are held at right angles and the first finger is pointed in the direction of the magnetic field and the thumb in the direction of motion then the second finger gives the direction of the induced current.

Lenz's law

Consider the energy changes that occur when a magnet is moved towards a coil. Assume that the magnet is moved towards the coil with its north pole facing towards the coil. Now by Lenz's law this induces a current in the coil such that the end of the coil nearest the magnet is an induced north pole. This repels the magnet and work must be done on the magnet to move it in against this repulsion.

The energy used produces the induced emf in the coil.

Eddy currents

These are induced currents in metal objects larger than pieces of wire. The emfs induced may not be very great but because the resistance of a lump of metal is low the induced currents can be large.

Since the induced currents always act so as to oppose the motion (Lenz's law) eddy currents can be used as a very effective electromagnetic brake. They also cause energy loss in transformers.

An emf generated in a moving straight wire in a magnetic field

A straight conductor of length L is moved through a field of flux density B. If the conductor moves with velocity v at right angles to the field then the flux cut per second will be BLv (since the conductor will sweep out an area vL every second). (See Figure 7.27.)

> **Revision activity**
>
> Find out about Michael Faraday's discovery of electromagnetic induction.

> **Lenz's law** states that the direction of the induced emf is such that it tends to oppose the change that produced it.

But the rate of cutting flux is equal to the emf induced in the conductor. Therefore:

emf generated $(\varepsilon) = BLv$

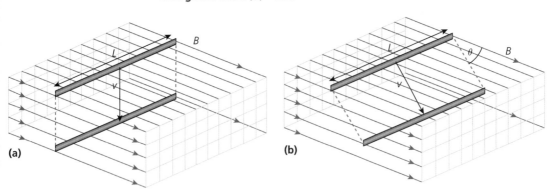

Figure 7.27 Induced emf in a straight wire

If the conductor cuts through the flux at an angle θ, where θ is the angle between the magnetic field and the direction of motion:

emf generated $(\varepsilon) = BLv \sin \theta$

The maximum emf is generated when the conductor moves at right angles to the field. $\theta = 90°$ and so $\sin \theta = \sin 90 = 1$.

Example

A wire of length 65 cm is moved at a constant speed of $4.5\,\text{m s}^{-1}$ through a magnetic field of flux density 0.2 T. If the wire is moving at right angles to the field calculate the emf generated between its ends.

Answer

emf generated $= BLv = 0.2 \times 0.65 \times 4.5 = 0.59\,\text{V}$

Now test yourself TESTED ☐

24 It is required to produce an emf of 10 V between the two ends of a 2 m long straight wire by moving it through a magnetic field of flux density 1.5 T at an angle of 25° to the field. Calculate the velocity of the wire needed.

Answer on p. 221

An emf generated in a rotating coil in a magnetic field

A coil of N turns and area A is rotated at a constant angular velocity ω in a magnetic field of flux density B, its axis being perpendicular to the field (Figure 7.28). When the normal to the coil is at an angle θ to the field the flux through the coil is $BAN\cos\theta = BAN\cos(\omega t)$, since $\theta = \omega t$.

The emf (ε) generated between the ends of the coil is:

emf generated $= BAN\omega\sin(\omega t)$

maximum emf, $\varepsilon_0 = BAN\omega$

Exam tip

The maximum value of the emf (ε_0) is when $\theta\,(= \omega t) = 90°$ (when the plane of the coil is in the plane of the field).

Exam practice answers and quick quizzes at **www.hoddereducation.co.uk/myrevisionnotes**

The emf is at its maximum value when the wires of the coil are cutting through the flux at right angles (Figure 7.28b). The root-mean-square value of the sinusoidal emf is:

$$E_{rms} = \frac{BAN\omega}{\sqrt{2}}$$

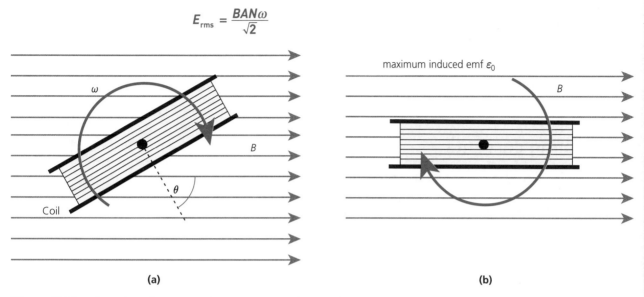

(a)   **(b)**

Figure 7.28 Induced emf in a rotating coil — varying emf

Figure 7.29 shows the induced emf at corresponding positions of the coil during its rotation.

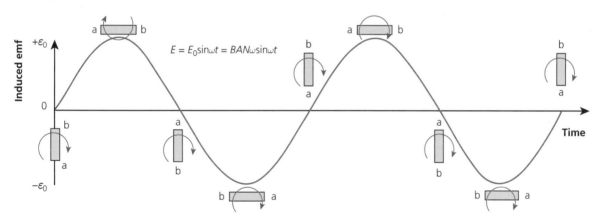

$$E = E_0 \sin\omega t = BAN\omega\sin\omega t$$

Figure 7.29 Coil positions and varying sinusoidal output voltage

---

Example

A circular coil of 100 turns, each of radius 10 cm, is rotated at 10 revs per second about an axis at right angles to a field of flux density 0.1 T. Find the position of the coil when the emf across its ends is a maximum, and calculate this emf.

Answer

The position of coil when the emf is maximum is when the plane of the coil is parallel to the field:

emf $= BAN\omega = 0.1 \times \pi \times 0.1^2 \times 100 \times 20\pi = 19.74$ V

Exam tip

Notice that the angular velocity used in the calculations must be in radians per second, where 1 rev s^{-1} = 2π rad s^{-1}.

Now test yourself

25 A circular coil of 200 turns, each of radius 8 cm, is rotated about an axis at right angles to a field of flux density 0.2 T.
  (a) Find the position of the coil when the emf across its ends is zero.
  (b) Calculate the required rate of revolution to give a maximum emf of 12 V between the ends of the coil.

Answer on p. 221

Alternating currents

Direct current is an electric current that flows in one direction only — the electrons drifting down the wire one way. If an **alternating voltage** is applied they change direction, first moving one way and then the other, and an **alternating current** is produced (Figure 7.30).

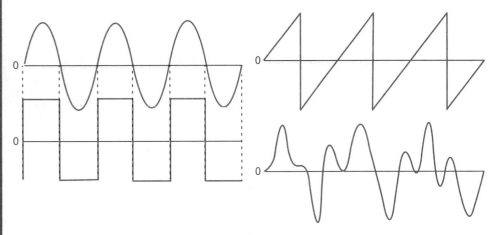

Figure 7.30 **Alternating voltage or current — variation with time**

An alternating current or voltage is one that varies with time about zero. We will restrict ourselves to a sinusoidal variation only (Figure 7.31).

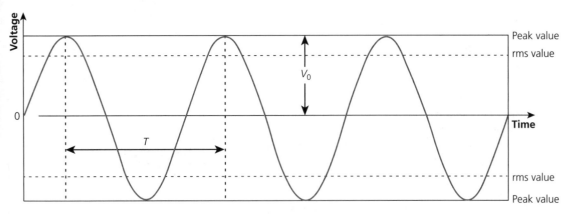

Figure 7.31 **Sinusoidal alternating voltage**

Root-mean-square values

The power P used in a resistor R is proportional to the square of the current:

$$P = I^2R$$

But with alternating current the value of I, and therefore of P, changes, and so:

mean value of P = (mean value of I^2) × R = I^2R

where:

$$I_{rms} = \sqrt{\text{mean value of } I^2} = \text{rms current}$$

For a sinusoidal variation of current and voltage the rms (root-mean-square) value of current I and voltage V are related to the peak value I_0 and v_0 by the following equations.

Root-mean-square current:

$$I_{rms} = \frac{I_0}{\sqrt{2}} = 0.707I_0$$

Root-mean-square voltage:

$$V_{rms} = \frac{V_0}{\sqrt{2}} = 0.707V_0$$

Example

In the UK the rms voltage of the mains electrical supply is 230 V.

Calculate the peak voltage.

Answer

$$V_{rms} = \frac{V_0}{\sqrt{2}} = 0.707V_0$$

So the peak value (V_0) is 230/0.707 or about 325 V.

Exam tip

Remember that the rms value is the square root of the mean-square value of the current or voltage.

Now test yourself

TESTED

26 A sinusoidal alternating pd has a peak value of 156 V. Calculate the rms value.
27 What is the mean value of a sinusoidal voltage?

Answers on p. 221

Use of an oscilloscope for measurement and display

An oscilloscope (Figure 7.32) has the following controls:
- on/off switch
- time base — this circuit applies a saw-tooth waveform to the X plates. The beam is moved from the left-hand side of the screen to the right during the time that the voltage rises to a maximum, and then is returned rapidly to the left as the voltage returns to zero.
- X input — voltage that controls the position of the beam in the x direction
- Y input — voltage that controls the position of the beam in the y direction
- Y gain — controls the sensitivity of the beam in the y direction (volts/cm)
- X gain — controls the sensitivity of the beam in the x direction (volts/cm)
- brightness — controls the brightness of the beam

An oscilloscope is a most useful instrument for the measurement and display of AC waveforms (Figures 7.32 and 7.33).

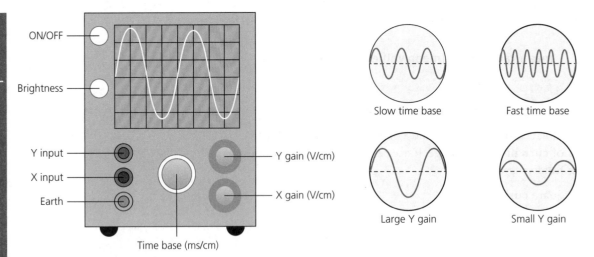

Figure 7.32 Oscilloscope and the waveforms displayed for a given AC input

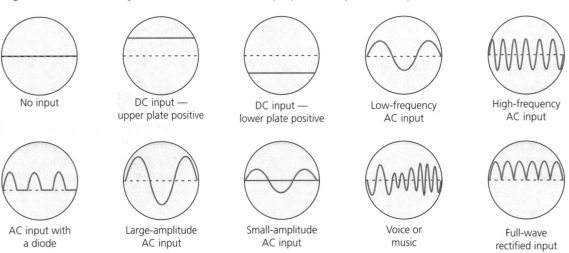

Figure 7.33 Oscilloscope displays for different inputs

Measuring voltage

Because of its effectively infinite resistance, the CRO makes an excellent voltmeter. It has a relatively low sensitivity, but this can be improved by the use of an internal voltage amplifier.

Most oscilloscopes have a previously calibrated screen giving the deflection sensitivity in volts per cm or volts per scale division. In this case a calibration by a DC source is unnecessary.

Measuring frequency

By using the calibrated time base, the input signal of unknown frequency can be 'frozen', and its frequency found directly by comparison with the scale divisions.

Alternatively, the internal time base can be switched off and a signal of known frequency applied to the X input. If the signal of unknown frequency is applied to the Y input, loops called Lissajous figures are formed on the screen. Analysis of the peaks on the two axes enables the unknown frequency to be found.

Now test yourself

28 Figure 7.34 shows a sinusoidal voltage displayed on an oscilloscope screen. If the grid shows cm squares, the time base has a sensitivity of $2\,ms\,cm^{-1}$ and the Y gain is $5\,V\,cm^{-1}$, determine:
   (a) the frequency of the signal
   (b) the amplitude of the signal
   (c) the value of the voltage at point A

Answer on p. 221

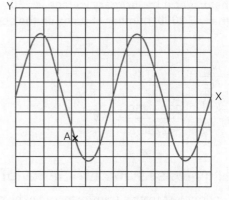

Figure 7.34 **Oscilloscope screen**

Eddy currents

Eddy currents are induced currents in metal objects larger than pieces of wire; the emfs induced may not be very great but because the resistance of a lump of metal is low, the induced currents can be large.

The operation of a transformer

REVISED ☐

In its simplest form a transformer consists of two coils known as the primary and secondary, wound on a laminated iron former that links both coils (Figure 7.35). The former, or core as it may be called, must be laminated otherwise large eddy currents would flow in it. The laminations are usually E–shaped, and the primary and secondary coils are wound one on top of the other to improve magnetic linkage, as in Figure 7.35b.

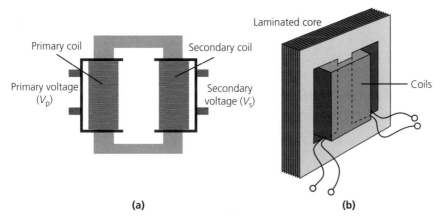

Figure 7.35 **The transformer**

An AC voltage is applied to the primary coil and this produces a changing magnetic field within it. This changing magnetic field links the secondary coil and therefore induces an emf in it. The magnitude of this induced emf (V_s) is related to the emf applied to the primary coil (V_p) by the equation:

$$\frac{V_s}{V_p} = \frac{N_s}{N_p}$$

where N_p and N_s are the number of turns on the primary and secondary coils respectively.

Example

Calculate the output (secondary) voltage of a 100% efficient transformer if the primary voltage is 230 V. (number of turns on the primary = 2000; number of turns on the secondary = 150)

Answer

$$\frac{V_s}{V_p} = \frac{N_s}{N_p}$$

$$V_s = \frac{V_p N_s}{N_p} = \frac{230 \times 150}{2200} = 16\,V$$

> **Exam tip**
>
> Remember that transformers will only work with alternating voltages.

Energy losses and efficiency of a transformer

The energy lost from a transformer may be due to:
- heating in the coils — this can be reduced by cooling the transformer, usually in oil
- eddy current losses in the core — reduced by the laminated core
- hysteresis loss — every time the direction of the magnetising field is changed some energy is lost due to heating as the magnetic domains in the core realign

Despite these energy losses transformers are remarkably efficient (up to 98% efficiency is common) and in fact among the most efficient machines ever developed.

$$\text{efficiency of a transformer} = \frac{I_s V_s}{I_p V_p}$$

If the transfer of energy from primary to secondary is 100% efficient:

$$I_s V_s = I_p V_p$$

Therefore:

$$\frac{V_p}{V_s} = \frac{N_p}{N_s}$$

Example

Calculate the output (secondary) voltage of an 85% efficient transformer if the primary voltage is 100 V, the primary current is 2 A and the secondary current 0.9 A.

Answer

$$\text{efficiency} = 0.85 = \frac{0.9 \times V}{100 \times 2}$$

$$V = \frac{200 \times 0.85}{0.9} = 189\,V$$

Now test yourself

TESTED

29 A certain transformer has 1000 windings on the primary coil and 2000 windings on the secondary coil. Calculate the output voltage when the following voltages are applied to the primary coil:
(a) a 12 V DC voltage
(b) a 25 V AC voltage
In each case assume that the transformer is 88% efficient.

Answer on p. 221

Transmission of electrical power

The transformer is a vital part of the National Grid, which distributes electrical energy around the country (Figure 7.36).

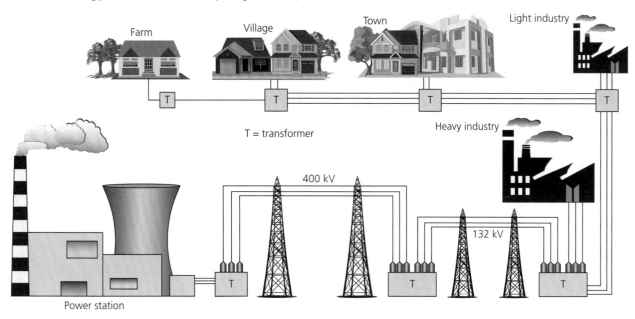

Figure 7.36 Transmission of electrical power

Electrical energy is generated in power stations by generators at a potential of 25 kV. It is then stepped up to 400 kV by a transformer and transmitted across the country in aluminium cables roughly 2 cm in diameter.

High voltages are used because the power loss (I^2R) per kilometre due to heating within the transmission cable, for a given power output, will be much less at high voltage and low current than at low voltage and high current.

Example

The resistance of a power cable is $0.1\,\Omega$ per km. Calculate the power loss (I^2R) per km for:

(a) a current of 1000 A

(b) when a 100:1 step-up transformer is used to reduce the current to 10 A

Answer

(a) power loss per km = 1000 × 1000 × 0.1 = 100 000 W

(b) power loss per km = 10 × 10 × 0.1 = 10 W

Now test yourself

TESTED ☐

30 Compare and contrast the use of aluminium or silver for electrical transmission lines.
  Density of aluminium = 2710 kg m⁻³; density of silver = 10 500 kg m⁻³
  Resistivity of aluminium = $3.21 \times 10^{-8}\,\Omega$ m; resistivity of silver = $1.6 \times 10^{-8}\,\Omega$ m

31 It is required to install transmission lines that give a maximum power loss per km of 8 W. If the current in the power lines is 10 A what is the resistance per km of these lines?

Answers on p. 221

Exam practice

1 (a) What is the unit for magnetic flux density? [1]
   (b) Define the unit that you have stated in part (a). [2]
   (c) A wire of length 0.6 m carrying a current of 2 A is placed at right angles to a magnetic field. If the force on the wire due to the magnetic field alone is 0.5 N calculate the strength of the magnetic field. [2]

2 (a) What is the unit for flux linkage? [1]
   (b) Define the unit that you have stated in part (a). [2]
   (c) What is the flux linkage through a coil of 50 turns and area of 25 cm² when it is placed at right angles to a field of flux density 0.5 T? [2]

3 An electron of mass 9×10^{-31} kg and charge 1.6×10^{-19} C moves in a circle of radius 10 mm in a magnetic field of flux density 6×10^{-4} T. Its speed in m s⁻¹ is:
   A 10^2 B 10^4 C 10^6 D 10^3 [1]

4 (a) A straight wire carrying a current *I* in the direction XY is placed between the poles of a magnet (Figure 7.37).

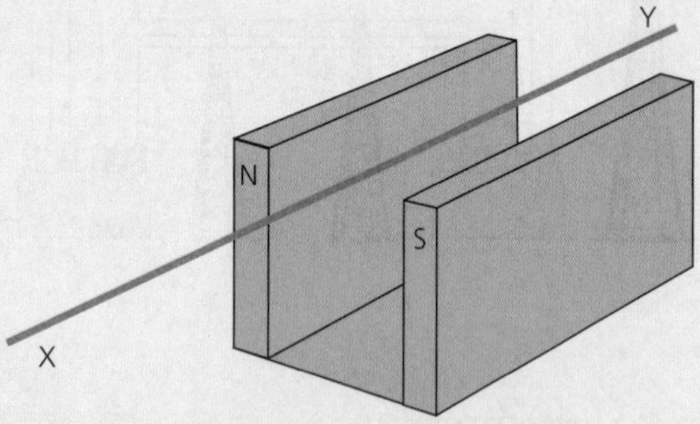

Figure 7.37 Magnet poles and a wire

   The resultant force on the wire is:
   A in the direction XY   B at right angles to both XY and NS
   C zero   D in the direction NS   [1]

5 A Boeing 747 with a wingspan of 60 m flies due south at a constant altitude in the northern hemisphere at 260 m s⁻¹. The vertical component of the Earth's magnetic field in that area is 4×10^{-5} T.
   (a) Calculate the emf between the wing tips and state which wing is positive. [2]
   (b) The aircraft now dives at 10° to the horizontal. Calculate the change in induced emf. (horizontal component of the Earth's magnetic field at this point = 2×10^{-5} T) [3]

6 A helicopter has a rotor with four blades each 6.4 m long and hovers in an area where the vertical component of the Earth's field is 4×10^{-5} T. If, as the rotor rotates, the tips of the rotor blades move with a speed of 200 m s⁻¹, calculate the induced emf:
   (a) between the tip of one blade and the axle [2]
   (b) between the tips of two diametrically opposite blades [1]
   (c) between the tips of two adjacent blades [1]

7 Describe and explain carefully what would happen to a transformer if the core was made of a single lump of metal with no laminations. [3]

8 In Japan electricity is transmitted at an rms voltage of 100 V and at 50 Hz.
   (a) What is the frequency of the supply? [1]
   (b) Calculate the peak voltage. [2]
   (c) A transformer with a step-down ratio of 2:1 is used to reduce the voltage. If the transformer is 85% efficient calculate:
      (i) the output rms voltage [2]
      (ii) the output current when an rms current of 2.3 A flows in the primary coil [3]

Answers and quick quiz 7 online

ONLINE

Summary

You should now have an understanding of:
- Electric fields — Coulomb's law and permittivity of free space
- Electric field strength — force per unit charge and field lines
- Electric potential — potential difference, and equipotential surfaces
- Capacitance — definition ($C = Q/V$); relative permittivity
- Parallel-plate capacitors — definition of permittivity
- Energy stored by a capacitor — calculation and area below the charge–pd graph
- Capacitance charge and discharge — graphical representation and calculation
- Magnetic fields — magnetic flux density
- Moving charges in magnetic fields — circular paths
- Magnetic flux and flux linkage — flux through a coil in a magnetic field
- Electromagnetic induction — Faraday's and Lenz's laws
- Alternating currents — peak and rms values of current and voltage
- The operation of a transformer — transformer equation, eddy currents, efficiency, transmission of electrical power

8 Nuclear physics

Radioactivity

Rutherford scattering

REVISED

The idea of atoms as small particles was put forward by the Greeks 2000 years ago, but the structure of the inside of the atom was not understood until the beginning of the twentieth century.

The English scientist Thomson suggested that the atom, which is a neutral particle, was made of positive charge with 'lumps' of negative charge inset in it — rather like the plums in a pudding. For this reason it was known as the plum pudding theory of the atom (Figure 8.1a).

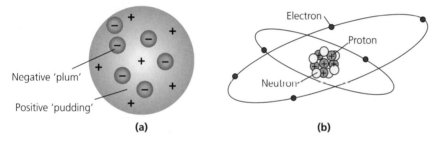

Figure 8.1 **Plum pudding and nuclear models**

The existence of a nucleus within an atom was shown by bombarding a thin gold foil with alpha particles in an experiment performed in 1911 by Rutherford, Geiger and Marsden. They were studying the passage of alpha particles through thin pieces of gold foil and found that some of the alpha particles were able to pass through the film. They then noticed something that they did not expect. Some of the alpha particles were being deflected from their original path and, more surprising, one in about 8000 were actually knocked backwards (Figure 8.2).

Rutherford knew that the alpha particles carried a positive charge so he said that the positive charge of the atom was concentrated in one place at the centre of the atom that he called the nucleus, and that the negatively charged particles, the electrons, were in orbit around the nucleus (Figure 8.1b).

> **Exam tip**
>
> The atom is mostly empty space.

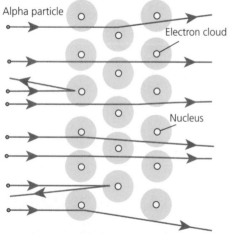

Figure 8.2 **Alpha particle scattering**

α, β and γ radiation

Nature of radioactive decay

In 1896, a previously unknown type of radiation was discovered. It was realised that this radiation:

- came from the nucleus of unstable atoms
- was spontaneous — once material containing these unstable nuclei was formed, emission could take place immediately and without the input of energy
- was random — it was impossible to predict which nucleus would emit radiation and when

The properties of α, β and γ radiation are outlined in Table 8.1.

Table 8.1

Property	Alpha (α)	Beta (β)	Gamma (γ)
Range in air	A few cm	Many cm	Metres
Ionising ability	Large	Small	Very small
Stopped by	Paper	mm of aluminium	cm of lead
Charge	$+2e$	$-e$	0
Mass	$8000m$	m	0
Deflection by a magnetic field	A little	A lot	None
Speed	$<1\%c$	$90\%c$	c

The energy of alpha particles emitted from a particular nucleus can have only one or two discrete values, dependent on the energy levels in the nucleus. The energy of emitted beta particles from a particular nucleus can have a range of values (Figure 8.3) because an antineutrino is also emitted (see page 16) and the two particles share the available energy.

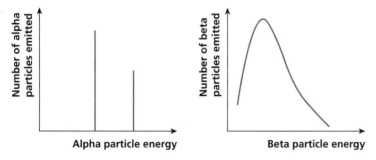

Figure 8.3 Alpha and beta radiation energy spread

Gamma radiation is emitted if the nucleus remains in an excited state after the alpha or beta emission.

Inverse square law for gamma radiation

The intensity of gamma radiation decreases with distance, and in a vacuum this decrease follows the inverse square law. In other words, if the distance (d) from the source is doubled, the intensity falls to one quarter ($\frac{1}{2}^2$) of the original value, to one ninth ($\frac{1}{3}^2$) if the distance is trebled and so on (Figure 8.4). In a material the decrease with distance is more rapid due to the interaction of the gamma radiation with the material.

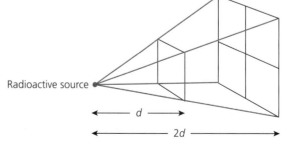

Figure 8.4 Decrease in the intensity of gamma radiation

This variation in intensity (I) is given by the equation:

$$I = \frac{k}{d^2}$$

Radiation in medicine

Although radiation is harmful to all cells in the body, the benefits of using it for radiotherapy and nuclear diagnosis often outweigh the dangers.

Therapy

The purpose of radiotherapy is to destroy malignant cancer cells without unduly affecting healthy tissue. This is achieved by rotating a beam of gamma radiation to three or four different positions around the body so that the paths of the beams cross at the tumour. All cells are affected but the fast-dividing malignant cells are affected more by the radiation than healthy cells.

Some tumours are treated by the implantation of a beta source within the tumour. The radiation emitted by the source is absorbed by the tumour.

Diagnosis

An isotope of technetium ($^{99}Tc^m$) is used as a source of gamma radiation for medical imaging and diagnosis. This has a short half-life (see page 178) of only 6 hours.

(see page 178)

> **Revision activity**
>
> Write a short account of the uses of radioisotopes in medicine with more detail than described here.

The difference between activity and count rate

The activity is the total number of emissions per second in all directions from the source. However, if you use a Geiger tube (or any other detector) to measure the number of emissions you will only record a fraction of the total emissions. This is because normal detectors do not usually surround the source and so you only detect particles (or photons) emitted into a small angle (Figure 8.5). This means that the observed count rate is always much less than the activity of the source.

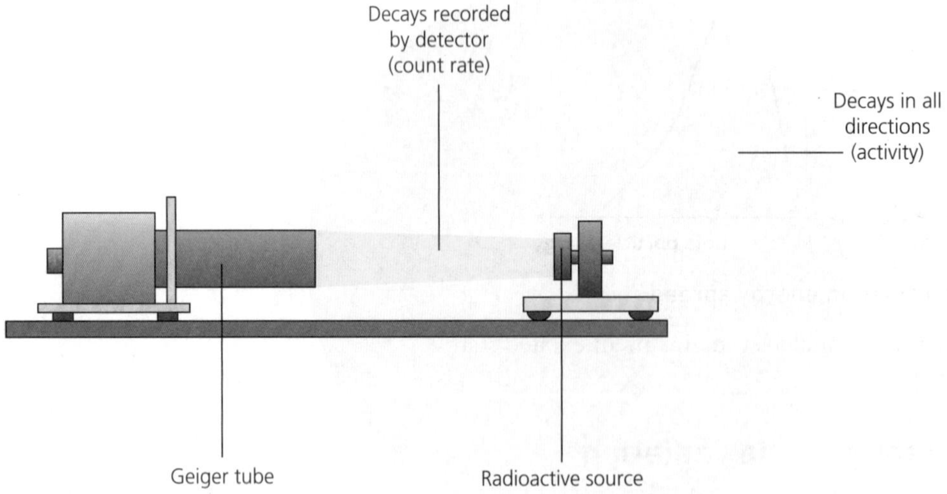

Decays recorded by detector (count rate)

Decays in all directions (activity)

Geiger tube

Radioactive source

Figure 8.5 Activity and count rate

Background radiation

A Geiger counter placed out in the lab well away from any radioactive sources will still read non-zero. It will probably record between 20 and

30 counts per minute. This **background radiation** is around us all the time (Figure 8.6).

Background radiation comes from:
- deep space — cosmic rays
- the Sun
- radioactive rocks such as granite
- radioactive material in our own bodies
- radon gas from the ground
- the nuclear industry, fallout from nuclear tests and medical uses

Radon gas

Radon is a naturally occurring gas that comes from the decay of uranium. Its short half-life makes it something we should be concerned about if it occurs in large quantities.

Concentrations of radon build up in underground caves and tunnels and can then seep into houses through minute cracks in hardcore and concrete floors to become trapped. Modern, well insulated houses are at more risk than older ones because they can act like sealed boxes (Figure 8.7). The danger is not great but it is something that people should be aware of.

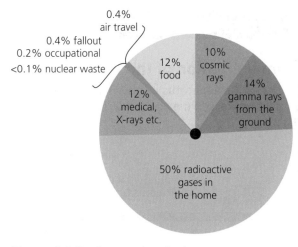

Figure 8.6 Background radiation

Exam tip

The count due to background radiation must be allowed for in experimental calculations such as that on page 180.

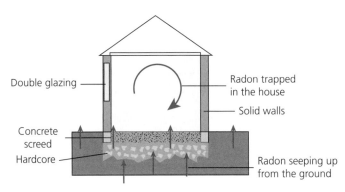

Figure 8.7 Radon gas in a house

Further (non-medical) uses of radioactive isotopes

These materials have a variety of uses, some of which are listed below:
- dating geological specimens using uranium, rubidium or bismuth
- dating archaeological specimens using carbon-14
- measuring paper, foil, steel sheet or plastic thickness using beta radiation
- sterilisation of foodstuffs
- liquid flow measurement
- tracing sewage or silt in the sea or rivers
- testing for leaks in pipes
- tracing phosphate fertilisers using phosphorus-32
- smoke alarms

Exam tip

The uses and hazards of radiation depend on the properties given in Table 8.1 — in particular, their relative ionising ability and hence range in air and penetration of materials.

Now test yourself

TESTED

1 Explain which of the following sources would be suitable for use in a home smoke detector.

$^{60}_{29}$Co   gamma emitter $^{241}_{95}$Am   alpha emitter $^{90}_{38}$Sr   beta emitter

Answers on p. 221

Required practical 12

Investigation of the inverse square law for gamma radiation

Set up the apparatus as shown in Figure 8.8, with the end window of the Geiger tube 10 cm from the source. Record the count rate for this distance.

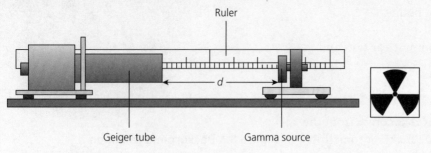

Figure 8.8 Inverse square law investigation

Move the Geiger away from the source and record the count rate every 2.5 cm.

Plot a graph of the count rate (C) against the inverse of the distance of separation ($1/d^2$).

Safety

A radioactivity warning notice should be displayed. Always use tongs to hold the radioactive source, and do not put your hands between the source and the Geiger counter. Follow all the normal school rules for the use and handling of radioactive material.

Radioactive decay

Mathematical treatment of radioactive decay

In a sample of radioactive material the number of nuclei (ΔN) decaying in a short time Δt is proportional to:

- N — the number of radioactive nuclei present at that moment
- Δt — the time over which the measurement is made
- the **decay constant** (λ) — a property of the element

$$\Delta N = -\lambda N \Delta t$$

The quantity $\Delta N/\Delta t$ is the rate of decay, or **activity** (A), of the source and is the number of disintegrations per second.

$$A = \frac{\Delta N}{\Delta t} = -\lambda N$$

The minus sign is there because the number of radioactive nuclei decreases as time increases.

The activity is measured in units called becquerels (Bq), where 1 Bq is one disintegration per second.

The decay constant (λ) can be defined as the probability of a nucleon decaying in the next second. Its unit is s^{-1}.

The number of radioactive nuclei (N) and the activity (A) of the source after a time t can be found using the following equations:

$$N = N_0 e^{-\lambda t}$$

$$A = A_0 e^{-\lambda t}$$

> **Exam tip**
>
> Remember that decay equations can be expressed in terms of the number of radioactive nuclei or in terms of the activity of the substance.

where N_0 is the initial number of nuclei present and A_0 is the initial activity of the source at time $t = 0$.

If we plot $\ln N$ against t we have a straight-line graph with gradient $-\lambda$ and an intercept on the $\ln N$ axis of $\ln N_0$ (Figure 8.9).

$$\ln N_0 - \ln N = \lambda t$$

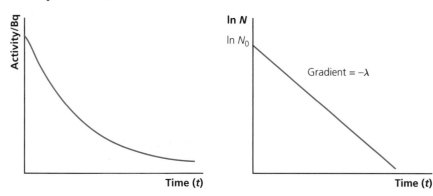

Figure 8.9 Radioactive decay

Example 1

The initial count rate of a sample of radioactive iodine-131 is 40 Bq. What will be the count rate after:
(a) 24 hours
(b) 2 weeks

(decay constant of iodine $= 1.43 \times 10^{-6}\,\text{s}^{-1}$)

Answer

(a) $A = A_0 e^{-\lambda t} = 40 \times e^{-(1.43 \times 10^{-6} \times 24 \times 60 \times 60)} = 35.4\,\text{Bq}$
(b) $A = A_0 e^{-\lambda t} = 40 \times e^{-(1.43 \times 10^{-6} \times 24 \times 60 \times 60 \times 14)} = 7.09\,\text{Bq}$

Example 2

The count rate due to carbon-14 in a sample of wood taken from an ancient Egyptian tomb was found to be 6.5 Bq when it was measured in 2016. If the count rate in living wood is 10 Bq how old is the archeological sample?

(decay constant of carbon-14 $= 3.95 \times 10^{-12}\,\text{s}^{-1}$)

Answer

$$6.5 = 10 \times e^{(-3.95 \times 10^{-12} \times t)}$$

$$\ln \frac{10}{6.5} = 3.95 \times 10^{-12} \times t$$

$$0.43 = 3.95 \times 10^{-12} \times t$$

Therefore:

$$t = \frac{0.357}{3.95 \times 10^{-12}} = 1.09 \times 10^{11}\,\text{s} = 3460\,\text{years}$$

Now test yourself

2 Calculate the activity of a sample of caesium-137 in the years:
   (a) 2020
   (b) 2035
   if the activity in 2016 was 185 kBq.
   (decay constant of caesium-137 = $7.85 \times 10^{-10}\,\text{s}^{-1}$)

3 A sample of radon gas escapes into a school laboratory. The decay constant of this alpha-emitting gas is $10\,\text{s}^{-1}$. How long would you need to wait before entering the lab if the safety regulations required the activity to have fallen to 10^{-6} of its original value?

Answers on p. 221

Half-life

As time passes the strength of a radioactive source gets weaker; its rate of decay ($\Delta N/\Delta t$) or activity gets smaller. The rate at which an unstable nucleus decays depends only on what type of nucleus it is. The decay is a random process. This means that if we take a sample of unstable nuclei (for example ^{226}Ra) we cannot know when any individual nucleus is going to decay. However we can measure what we call the **half-life** ($T_{\frac{1}{2}}$) for the element in the sample. This is constant (Figure 8.10) and defined as:

● The average time taken for half the original number of nuclei in a sample of an element to decay.

or:

● The average time taken for the activity of a radioactive source to fall to one half of its original value.

Typical mistake

Not using the correct units (s^{-1}) for the decay constant.

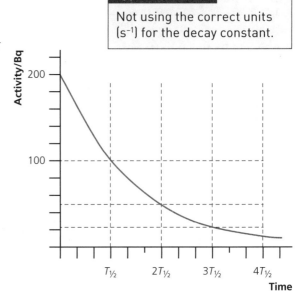

Figure 8.10 Half-life

Example

The activity of a sample of radioactive material with a half-life of 2 minutes is 384 Bq at the start. What will be the activity after:
(a) 2 minutes   (b) 4 minutes
(c) 8 minutes   (d) 16 minutes?

Answer

(a) One half-life   activity = 192
(b) Two half-lives   activity = 96
(c) Four half-lives   activity = 24
(d) Eight half-lives   activity = 1.5

Exam tip

Remember that the activity halves every half-life and not when the time passed doubles.

Half-life and the decay constant

Half-life is related to the decay constant by the equation:

$$T_{1/2} = \frac{\ln 2}{\lambda}$$

λ can be replaced by $(\ln 2)/T_{\frac{1}{2}}$ in calculations, where appropriate.

Example 1

A laboratory buys a sample of radioactive material that has a half-life of 150 years. If the initial activity is 200 Bq what will be the activity after 10 years?

Answer

$$A = A_0e^{-\ln 2 \, t/T_{1/2}} = 200e^{-\ln 2 \times 10/150} = 200e^{-0.046} = 200 \times 0.96 = 191\,\text{Bq}$$

Exam tip

In Example 1, the time and the half-life can be kept in years because they appear in the formula as a ratio.

Example 2

The decay constant (λ) of a particular isotope (radon-220) is $1.33 \times 10^{-2}\,\text{s}^{-1}$. How long will it take for the activity of a sample of this isotope to decay to one eighth of its original value?

Answer

$$\text{half life} = \frac{0.693}{\lambda} = \frac{0.693}{1.33 \times 10^{-2}} = 52\,\text{s}$$

number of half lives required to reduce the activity to one eighth = 3

Therefore:

time needed $= 3 \times 52 = 156\,\text{s}$

A useful alternative formula for calculating final activity is:

$$A = \frac{A_0}{2^n}$$

where n is the number of half-lives that have passed.

The formula works not only for simple cases where n is a whole number (i.e. for one half-life, two half-lives etc.) but also when n is any number (i.e. 1.2 half-lives, 4.3 half-lives and so on).

Radioactive decay, molar mass and the Avogadro constant

The number of nuclei in a sample can be related to the mass of the source, using the molar mass and Avogadro's number, by the formula:

$$\text{mass } (m) = \frac{\text{molar mass } (M) \times \text{number of nuclei } (N)}{\text{Avogadro constant } (N_A)}$$

This formula can be used use it to find out the mass of a given source if we know its activity.

Example

A school has a radium-226 source with an activity of 185 000 Bq. What is the mass of the source?

(Avogadro constant $(N_A) = 6.02 \times 10^{23}$ mol^{-1}; decay constant for radium-226 $= 1.35 \times 10^{-11}$ s^{-1})

Answer

Using the formula:

$$A = -\lambda N = -\lambda \frac{m}{M} N_A$$

$$185\,000 = 1.35 \times 10^{-11} \frac{m}{226} \times 6.02 \times 10^{23}$$

Therefore:

$$m = \frac{185\,000 \times 226}{1.35 \times 10^{-11} \times 6.02 \times 10^{23}} = 5.14 \times 10^{-6}\,\text{g} = 5.14\,\mu\text{g}$$

Now test yourself

TESTED

4 A scientist measured the activity due to caesium-137 contamination in a field near to the Chernobyl power station just after the accident in 1986 and found it to be 10 kBq m^{-2}. If the half-life of caesium-137 is 30 years, what would be the activity in:
   (a) 2016   (b) 2026?
5 A radioactive sample decays with a half-life of 10 hours. If the background count is 10 Bq and the original reading of the Geiger counter is 170 Bq, what is the reading after:
   (a) 20 hours   (b) 25 hours   (c) 30 hours?
6 A sample of radioactive fluorine gave a measured count rate of 45 Bq when measured in a laboratory where the background count is 2 Bq. What will be the measured count rate after:
   (a) 4 hours   (b) 12 hours
   Comment on your answer to (b). The half-life of radioactive fluorine is 1.8 hours.

Answers on p. 222

Nuclear instability

REVISED

Variation of N and Z for stable nuclei

The variation in proton and neutron numbers is a very important factor in our understanding of the behaviour of nuclei. The graph in Figure 8.11 shows the variation of neutron number with proton number for stable nuclei. For light atoms (such as helium) these two numbers are equal — there are as many neutrons in the nucleus as there are protons. However, for more massive elements, the neutron number is much greater than the proton number. For example, a $^{235}_{92}$U nucleus contains 92 protons but 143 neutrons.

The position of an unstable isotope relative to the stability line determines whether the nucleus will decay by β^+, β^- or α emission.

Exam practice answers and quick quizzes at **www.hoddereducation.co.uk/myrevisionnotes**

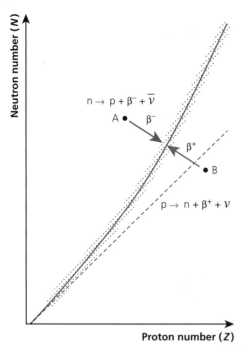

Figure 8.11 Graph of N against Z for stable nuclei

Changes in N and Z caused by radioactive decay

Radioactive decay equations

Unstable nuclei may lose some of their excess energy by a variety of radioactive processes. Examples of α, β and γ are shown by the following equations and in Figure 8.12.

Alpha emission:

$$^{226}_{88}\text{Ra} \rightarrow {}^{222}_{86}\text{Rn} + {}^{4}_{2}\text{He}$$

Beta emission:

$$^{90}_{38}\text{Sr} \rightarrow {}^{90}_{39}\text{Y} + {}^{0}_{-1}\text{e} + {}^{0}_{0}\bar{v}$$

Gamma emission:

$$^{60}_{27}\text{Co} \rightarrow {}^{60}_{27}\text{Co} + {}^{0}_{0}\gamma$$

Gamma ray emission is evidence for excited states within the nucleus.

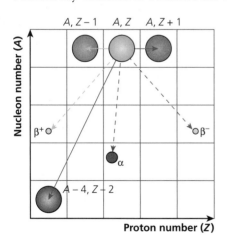
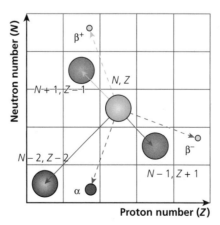

Figure 8.12 Alpha and beta emission grids

Now test yourself

7 In the nuclear reaction $^{22}_{11}Na + p \rightarrow {^A_Z}X + n$ the proton number Z and the nucleon number A of the nucleus X are:

	Proton number Z	Nucleon number A
A	12	23
B	11	22
C	12	22
D	10	23

8 All the decays in the uranium decay series occur by the emission of just one particle for each change (sometimes a gamma ray is emitted as well). Write the name of the particle (alpha or beta) that is emitted at each stage.

Isotope	Half-life	Particle emitted
Polonium-218	3.1 minutes	
Lead-214	27 minutes	
Bismuth-214	20 minutes	
Polonium-214	1.6×10^{-4} s	
Lead-210	19 years	
Bismuth-210	5.0 days	
Polonium-210	138 days	
Lead-206	stable	

Answers on p. 222

Electron capture

An unstable atom may also become stable by a process called **electron capture**. An electron from one of the atom's inner orbits is drawn into the nucleus, where it combines with a proton to form a neutron. One possible nuclear equation for this process is:

$$^{11}_{6}C + {^0_{-1}}e \rightarrow {^{11}_{5}}B + {^0_0}\nu$$

Excited states within the atomic nucleus

Energy emission in the form of gamma radiation occurs due to transitions between excited states within the nucleus. This emission is very good evidence for the existence of such states.

Revision activity

Make a mind map to contrast and compare the nature and properties of alpha, beta and gamma radiation.

Nuclear radius

REVISED

The radius of the atomic nucleus

Fermi proposed the following equation for the radius of a nucleus (r) in terms of the nucleon number of the nucleus (A):

$$r = r_0 A^{1/3}$$

where the constant r_0, the radius of the hydrogen nucleus, is 1×10^{-15} m = 1 fm. Since $r^3 = r_0^3 A$, nuclear volume is proportional to nuclear mass and so nuclear matter has a constant density.

Exam practice answers and quick quizzes at **www.hoddereducation.co.uk/myrevisionnotes**

Example

Calculate the radii of the following nuclei:
(a) carbon-12
(b) gold-197

Answer

(a) $r = 10^{-15} \times 12^{1/3} = 2.3 \times 10^{-15}$ m $= 2.3$ fm
(b) $r = 10^{-15} \times 197^{1/3} = 5.8 \times 10^{-15}$ m $= 5.8$ fm

Distance of closest approach

If an alpha particle (charge $+2e$) with a kinetic energy E_α is fired directly towards a nucleus of change Q it will 'feel' a repulsion, which increases as it gets closer — climbing the potential 'hill' surrounding the nucleus. When all the kinetic energy has been converted to electric potential energy (see p. 146) the alpha particle (charge q) has reached its distance of closest approach (d_c) and so comes to rest before moving away from the nucleus.

$$E_\alpha = 2eV = \left(\frac{1}{4\pi\varepsilon_0}\right)\left(\frac{2eQ}{d_c}\right)$$

Therefore:

$$\text{distance of closest approach } (d_c) = \left(\frac{1}{4\pi\varepsilon_0}\right)\left(\frac{2eQ}{E_\alpha}\right)$$

Now test yourself

TESTED

9 Calculate the distance of closest approach between an alpha particle ($Z = 2$) of energy 4.5 MeV and a gold nucleus ($Z = 79$). ($\varepsilon_0 = 8.85 \times 10^{-12}$ Fm^{-1} and $e = 1.6 \times 10^{-19}$ C)

Answer on p. 222

Calculation of nuclear density

The simplest atomic nucleus is that of hydrogen (one proton) with a radius of the order of 10^{-15} m. The mass of a proton is about 10^{-27} kg and this gives a density for the nucleus of the order of 10^{17} kg m^{-3}.

Typical mistake

Using MeV and not joules in nuclear physics calculations.

Electron diffraction by a nucleus

From the Fermi equation ($r = r_0 A^{1/3}$) we know that the 'diameter' of a nucleon is about 10^{-15} m and so if we want to have any chance of 'seeing' inside a nucleus we need to use a wavelength of this order or even smaller.

Electrons have wave as well as particle properties, so they can be diffracted. De Broglie's equation (wavelength $(\lambda) = h/mv = hc/E$, where h is the Planck constant and E the energy of the electrons) can be used to calculate their wavelength, which for 300 MeV electrons is 4.1×10^{-15} m. This is comparable to the diameter of nuclei and so electron diffraction can be used to determine the size of a nucleus (Figure 8.13). (See also Figure 2.27 on page 30.)

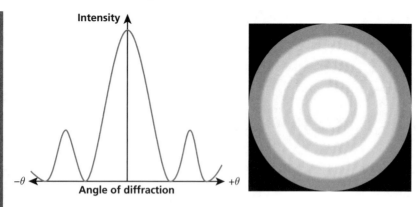

Figure 8.13 Electron diffraction by a nucleus

Example

A 380 MeV electron is diffracted through an angle of 40° by a nucleus. Calculate the diameter of the nucleus (d).
Angle of diffraction (θ) by a circular obstacle of diameter d is given by:

$$\sin \theta = 1.22\ \lambda/d$$

Answer

$$\text{electron wavelength} = \frac{hc}{E} = \frac{6.63 \times 10^{-34} \times 3 \times 10^{8}}{380 \times 10^{6} \times 1.6 \times 10^{-19}} = 3.11 \times 10^{-15}\,\text{m}$$

angle of diffraction (θ) = 40°

$$\sin 40 = 0.643 = \frac{1.22\lambda}{d}$$

$$d = \frac{1.22\lambda}{0.643} = \frac{1.22 \times 3.11 \times 10^{-15}}{0.643} = 5.90 \times 10^{-15}\,\text{m}$$

Mass and energy

REVISED

Einstein's mass–energy equation

Matter and energy are related by the famous equation proposed by Einstein:

$$\Delta E = \Delta mc^2$$

The quantity ΔE is the amount of energy produced when a mass Δm is completely converted to energy, and c is the speed of light ($3 \times 10^8\,\text{m s}^{-1}$).

Example

Calculate the energy produced when 150 g of any matter is converted into energy.

Answer

By Einstein's equation:

$$\Delta E = \Delta mc^2$$

$$\Delta E = 0.15 \times [3 \times 10^8]^2 = 1.35 \times 10^{16}\,\text{J}$$

Exam tip

Note the use of kg and not g in the calculation.

Atomic mass unit (u)

The masses of atoms and subatomic particles are extremely small — for example, the mass of an oxygen atom is about 3×10^{-26} kg and that of a neutron about 1.67×10^{-27} kg. It is therefore convenient to define a new unit to measure them. This is known as the **atomic mass unit** — written as u.

One atomic mass unit is defined as one twelfth of the mass of one atom of the carbon-12 isotope.

$$1\,u = \frac{1}{12}(19.92 \times 10^{-27})\,kg = 1.661 \times 10^{-27}\,kg$$

Alternative units for energy and mass

The MeV (see p.28) is related to the joule by:

$$1\,MeV = 10^6\,eV = 1.6 \times 10^{-13}\,J$$

But using Einstein's mass–energy relationship ($E = mc^2$) we can convert the units for mass (kg) into alternative units. The units for mass are MeV/c^2. Therefore:

$$1\,u = 1.66 \times 10^{-27}\,kg = 931.5\,MeV/c^2$$

The proton therefore has a rest *mass* of 1.007273 u or 938 MeV/c^2 and a rest *energy* of 938 MeV.

Nuclear binding energy

Imagine that you were asked to make a nucleus. You are given the protons and neutrons, asked to measure their masses, make them stick together somehow and then measure the mass of the finished nucleus. You would find that the mass of the completed nucleus is less than the total mass of the protons and neutrons from which it was made.

The difference in mass between the mass of the nucleus and that of the particles of which it is composed is called the **mass defect** of the nucleus. The larger the nucleus the larger the mass defect.

The mass defect can be expressed as an energy, and this is called the **binding energy** of the nucleus (Figure 8.14).

The binding energies and mass defects for a number of nuclei are shown in Table 8.2.

> **Exam tip**
>
> Sometimes masses are given in MeV and not MeV/c^2.

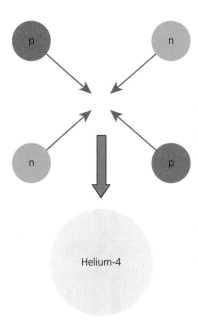

Figure 8.14 Nuclear binding energy

Table 8.2

Element	Mass defect/u	Binding energy/MeV
$^{12}_{6}$C (carbon)	0.09900	92.2
$^{16}_{8}$O (oxygen)	0.13708	127.6
$^{40}_{20}$Ca (calcium)	0.36741	342.2
$^{56}_{26}$Fe (iron)	0.52875	491.8
$^{208}_{82}$Pb (lead)	1.75784	1637.4
$^{235}_{92}$U (uranium)	1.93538	1802.8

Example

Calculate the mass defect and binding energy of the helium-4 nucleus.

Answer

mass of two protons = $2 \times 1.00728\,u = 2.01456\,u$

mass of two neutrons = $2 \times 1.00867\,u = 2.01734\,u$

total mass of the particles = $4.0319\,u$

mass of the helium-4 nucleus = $4.00151\,u$

mass defect = $0.03039\,u$

binding energy of the helium-4 nucleus = $0.03039 \times 931.5 = 28.3\,MeV$

Binding energy per nucleon

Another useful quantity is the **binding energy per nucleon**. It is defined as:

$$\text{binding energy per nucleon} = \frac{\text{binding energy}}{\text{nucleon number}}$$

Figure 8.15 shows the binding energy per nucleon against nucleon number. Elements with a high binding energy per nucleon are very difficult to break up. Iron-56 is close to the peak of the curve and has one of the highest binding energies per nucleon of any isotope.

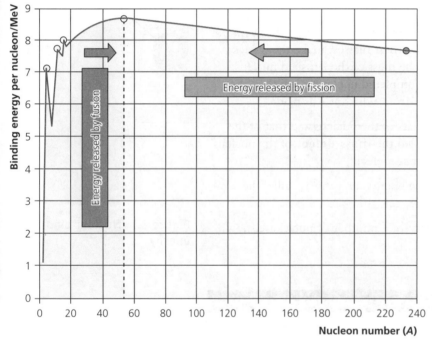

Figure 8.15 Binding energy per nucleon

The part of the curve to the left shows that two light elements can produce energy by fusion, while the part of the curve to the right shows that a heavy element can produce energy by fission. Notice that the diagram has been drawn with the binding energy per nucleon being shown as positive — this represents the energy needed to separate the particles.

> **Example**
>
> The mass of the isotope $^{7}_{3}$Li is 7.016 u. Find its binding energy.
>
> *Answer*
>
> Protons: $3 \times 1.007276 = 3.021829$
>
> Neutrons: $4 \times 1.008665 = 4.03466$
>
> Total $= 7.056489$
>
> Nucleus $= 7.016$
>
> Mass defect $= 0.040489$ u
>
> Binding energy $= 0.040489 \times 931.5 = 37.70$ MeV

Now test yourself

TESTED

10 The isotope of iron $^{57}_{26}$Fe has a mass of 56.935 u. Calculate:
    (a) the binding energy of the nucleus of $^{57}_{26}$Fe
    (b) the binding energy per nucleon for $^{57}_{26}$Fe
    (mass of a proton = 1.00728 u; mass of a neutron = 1.00867 u; 1 u = 1.661×10^{-27} kg = 931.5 MeV)

Answer on p. 222

Nuclear fission

REVISED

Nuclear fission is the splitting of a heavy nucleus by the bombardment of this nucleus by smaller particles — usually neutrons.

There are many possible results of the nuclear fission of an isotope of uranium, uranium-235. One possible reaction is:

$$^{235}_{92}U + ^{1}_{0}n \rightarrow ^{236}_{92}U \rightarrow ^{148}_{57}La + ^{85}_{35}Br + 3^{1}_{0}n + energy$$

$235.044 + 1.0087 = 236.0527$ u $147.961 + 84.938 + 3.0261 = 235.9251$ u

This reaction has a mass defect of 0.1276 u.

Energy is given out by the reaction because the mass of the products is less than the total mass of the original nucleus and the neutron.

The mass defect represents a very small amount of energy. However, when you work out how many nuclei there are in 1 kg of uranium you can understand why nuclear fission is so important.

Now test yourself

TESTED

11 If the mass defect in the fission of one nucleus of uranium-235 is 0.1276 u, calculate the energy in joules available by the fission of 1 kg of uranium-235. (Avogadro's number = 6.02×10^{23} mol^{-1})

Answer on p. 222

Fission and fusion processes

Induced fission by thermal neutrons

When a thermal neutron is fired at a uranium-235 nucleus, uranium-236 is formed. The uranium-236 nucleus is unstable and may lose the extra energy in two ways. It can emit radiation (alpha, beta or gamma) or break apart. This 'breaking up' of the nucleus was the first evidence of nuclear fission (Figure 8.16).

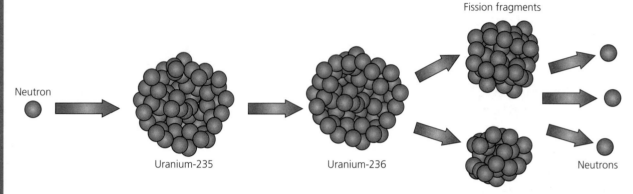

Neutron Uranium-235 Uranium-236 Fission fragments Neutrons

Figure 8.16 Induced nuclear fission

Thermal neutrons are ones with velocities of the order of $10^4 \, \text{m s}^{-1}$. These relatively low velocities allow time for the nucleus to 'capture' the incoming neutron, so forming the unstable nucleus.

Chain reaction and critical mass

Once one nucleus has undergone fission the neutrons that are released can go on to split further nuclei. If this fission can be sustained a **chain reaction** is produced (Figure 8.17). This reaction will proceed at high speed; the time for an emitted neutron to collide with another nucleus to produce a second fission is about $0.01 \, \mu\text{s}$.

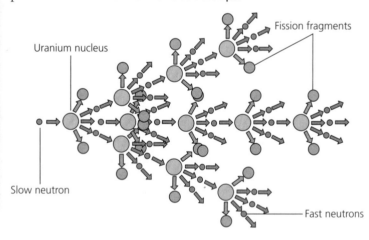

Uranium nucleus Fission fragments Slow neutron Fast neutrons

Figure 8.17 Chain reaction

12 Calculate the energy available from the fission of 1 kg of plutonium-239 by the reaction shown below. Give your answer in joules.

$$^{239}_{94}\text{Pu} + ^{1}_{0}\text{n} \rightarrow ^{134}_{54}\text{Xe} + ^{103}_{40}\text{Zr} + 3^{1}_{0}\text{n}$$

(Additional nuclear mass data: $^{239}\text{Pu} = 239.060765 \, \text{u}$, $^{134}\text{Xe} = 133.9054 \, \text{u}$, $^{103}\text{Zr} = 102.9266 \, \text{u}$, $\text{n} = 1.00867 \, \text{u}$)

Answer on p. 222

Thermal nuclear reactors

A nuclear reactor needs a minimum amount of fuel called the **critical mass** (about 15 kg for pure uranium-235) to sustain a chain reaction. Anything less than this and the loss of neutrons from the surface will be too great and the chain reaction will stop.

Important parts of a thermal nuclear reactor (Figure 8.18) include the following:

- **Fuel** — in most commercial thermal nuclear reactors this is usually uranium-235 or 238, often a mixture.
- **Moderator** — the neutrons produced in a chain reaction are moving too fast to cause further fission in ^{235}U nuclei and they have to be slowed down. Moderators used are graphite or heavy water.
- **Control rods** — these are designed to control the rate of the reaction by absorbing neutrons without undergoing nuclear fission. Lowering them into the reactor core will slow down the reaction. Rods of boron steel are used.
- **Coolant** — a liquid or a gas used to remove the heat energy from the reactor core and keep its temperature stable. Coolants may be either carbon dioxide or water.
- **Steel containment vessel** — a thick, steel vessel to contain the high-pressure gas coolant.
- **Concrete biological shield** — 2–3 m thick, to protect the operators from radiation.

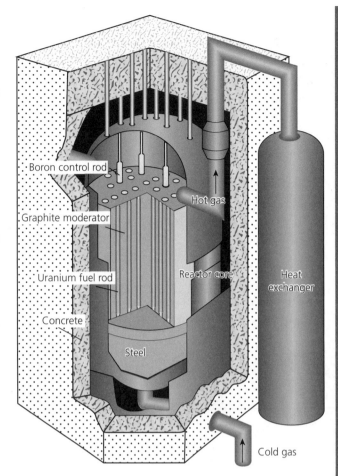

Figure 8.18 Thermal nuclear reactor

Safety aspects

There are a number of considerations and problems associated with the commercial use and decommissioning of nuclear fission reactors. Considerations when choosing the site include:

- transport facilities
- environmental aspects
- local industry and commerce
- a hard-rock site
- local centres of population
- availability of cooling water

For safety, any overheating in the reactor leads to automatic shut-down. The control rods are held by electromagnets. If the temperature in the core rises to an unsafe level the current in these magnets is switched off and the control rods fall into the core, absorbing most of the neutrons and so shutting off the reaction.

Disposal of nuclear waste

In the UK, low-level waste, such as gloves, cast-off clothing and over shoes, is encased in cement and stored 'on site' for up to 15 years. After that it is packaged and disposed of as 'normal' waste.

The intermediate-level waste with long half-lives, such as fuel containers, is packed in 500-litre steel drums. These are currently stored (about 50 000 m^3 so far).

The most radioactive is the high-level waste, typically 1000 times more radioactive than intermediate level waste, and is mainly 'used' fuel elements taken from reactors. In future, high-level waste from decommissioning will be stored in 12 m^3 steel boxes, with the spaces between items packed with concrete.

Decommissioning a nuclear reactor

Although nuclear power is a very good way of producing energy, there are problems when the reactor reaches the end of its useful life.

After removal from the reactor 'used' fuel elements are stored under water in 'ponds' while they cool. Fuel is reprocessed to take out the uranium and plutonium but there is still some very toxic waste left behind, which is then encased in glass blocks. The government has decided to store this material underground for 50 years before disposal.

Nuclear power and the greenhouse effect

It has been calculated that if the use of nuclear power were expanded, CO_2 emissions could be reduced by up to 30%, thus lowering global warming by 15%.

Nuclear fusion

REVISED

If two light nuclei can be joined together we have another way of releasing energy — this is known as **nuclear fusion** (Figure 8.19).

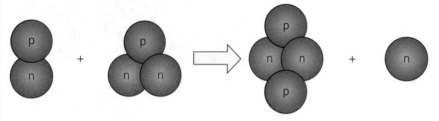

Figure 8.19 Nuclear fusion

The most likely fusion reaction is the deuterium (D)–tritium (T) one shown below:

$$^2_1D + ^3_1T \rightarrow ^4_2He + ^1_0n + \text{energy (17.6\,MeV)}$$

2.014 102 + 3.016 049   4.002 604 + 1.008 665

This gives a mass defect of 0.018 888 u and available energy of 17.6 MeV. This is less than for the fission of one nucleus of uranium, but since the density of deuterium is so much less than that of uranium the yield per kilogram is comparable. In 1 kg of deuterium there are approximately 3×10^{26} atoms and so the energy released per kilogram is 8.45×10^{14} J.

The big problem here is that both the nuclei are positive and to get them to fuse we have to somehow make them come very close together, so that the strong nuclear force becomes greater than the electrostatic repulsion. The way this is done is to make them collide at very high speed by raising the temperature of the gas to over 100 million °C — several times hotter than the centre of the Sun. At these temperatures the gas becomes a plasma — a sea of electrons and ions — and is a real problem to contain.

Now test yourself

TESTED

13 Suggest why the critical ignition temperature is higher for a D–D reaction than for a D–T reaction.
14 Calculate the energy released in the following reactions:
   (a) $D + D \rightarrow n + {}^3He$ (b) $D + D \rightarrow {}^3H + {}^1H$ (c) $T + T \rightarrow 2n + {}^4He$
   (Additional nuclear mass data: D = 2.014 102 u; ${}^3He$ = 3.016 048 u; ${}^3H$ = T = 3.016 049 u; n = 1.00867 u)

Answers on p. 222

Exam practice

1 One conclusion from the Rutherford scattering experiment is:
   A   Alpha particles are helium nuclei.
   B   Electrons do not interact with neutrons.
   C   The mass and positive charge in a gold atom is concentrated in a nucleus.
   D   Alpha particles cannot pass through gold foil.   [1]

2 A Geiger counter with an end window of diameter 1.5 cm is placed 20 cm from a cobalt-60 gamma source containing 1.3×10^8 radioactive nuclei. The counter records a count rate of 200 Bq.
   (a) What is the activity of the source?   [2]
   (b) What is the decay constant for cobalt-60 (using the data from this question)?
       Ignore background count in this question.   [2]

3 The decay constant (λ) can be written in terms of the half-life $(T_{\frac{1}{2}})$ as:

   A $\ln 2 \times T_{\frac{1}{2}}$ B $\dfrac{T_{\frac{1}{2}}}{\ln 2}$ C $\dfrac{\ln 2}{T_{\frac{1}{2}}}$ D $\ln 2 - T_{\frac{1}{2}}$ [1]

4 A sample of wood from an archaeological site is carbon dated. The activity is found to be 600 counts per gram per hour. If the initial count rate of living wood is 950 counts per gram per hour how old is the wood sample? (Half-life of carbon-14 = 5570 years; background count has been allowed for.) [3]

5 A laboratory prepares a 2 µg sample of caesium-137, with half-life of 28 years.
   (a) What is its initial activity?   [2]
   (b) What will its activity be after 100 years?   [2]
   (c) What is the decay constant for caesium (in s^{-1})?   [2]
   (Avogadro constant = 6.02×10^{23})

6 A patient was given an injection containing a small amount of the isotope sodium-24, which is a beta emitter with a half-life of 15 hours. The initial activity of the sample was 60 Bq. After a period of 8 hours the activity of a 10 ml sample of blood was found to be 0.08 Bq.
   (a) Estimate the volume of the patient's blood from these measurements.   [3]
   (b) What assumptions have you made in your calculation?   [2]

7 Show that the binding energy of $^{58}_{28}\text{Ni}$ is approximately 492 MeV. (mass of Ni-58 nucleus = 57.9353 u) [3]

8 Using the binding energy per nucleon graph (Figure 8.15), which of the following gives the best explanation of why energy can be released in nuclear fission?
   A   The graph slopes upwards as the nucleon number increases from 1 to 50.
   B   The graph slopes downwards as the nucleon number increases from 60 to 250.
   C   Part of the graph has a steep gradient.
   D   Part of the graph has a shallow gradient.   [1]

9 Explain the difference between the functions of the moderator and the control rods in the operation of a nuclear fission reactor.   [1]

10 (a) Suggest why the shape of the sample of nuclear fuel is important in sustaining a nuclear chain reaction.   [2]
   (b) The carbon dioxide codant gas in a nuclear reactor circulates at a pressure of about 4×10^6 Pa. This requires a very strong steel containment vessel surrounding the reactor core. Why is such a high pressure needed and what are the resulting requirement for the reactor design? [2]
   (c) If the core temperature of a nuclear fission reactor is 300 K calculate the root mean square speed of thermal neutrons within the core.   [2]
   (mass of a neutron = 1.67×10^{-27} kg; Boltzmann constant = 1.38×10^{-23} J K^{-1})

Answers and quick quiz 8 online

ONLINE

Summary

You should now have an understanding of:
- Rutherford scattering — alpha particles scattered by the very small positive atomic nucleus
- Alpha, beta and gamma radiation — their nature and properties
- Radioactive decay — random nature, decay constant, half-life, applications of radioactive decay
- Nuclear instability — graph of neutron number number against proton number

- Nuclear radius — closest approach in alpha particle scattering, dependence of nuclear radius on nucleon number, nuclear density, diffraction of electrons by nuclei
- Mass and energy — fission and fusion processes
- Induced fission — chain reaction, critical mass, functions of parts of a nuclear reactor
- Safety aspects and decommissioning in nuclear power production

9 Astrophysics

Telescopes

Astronomical refracting telescope

The main purposes of a telescope used for astronomy are:
- to gather as much light as possible — this is done by using a large aperture lens or mirror. The amount of light gathered depends on the *area* of the lens or mirror.
- to resolve fine detail — this is also done by using a large aperture lens or mirror. The larger the aperture the finer the detail that can be seen.
- to magnify the image of a distant object — this is done by using a lens or mirror with a long focal length.

> **Exam tip**
>
> The power of a lens is $1/f$ where f is the focal length in metres.

Astronomical refracting telescope in normal adjustment

Figure 9.1 shows an astronomical refracting telescope. A simple astronomical refractor is usually adjusted so that the final image is at infinity. This is known as **normal adjustment**.

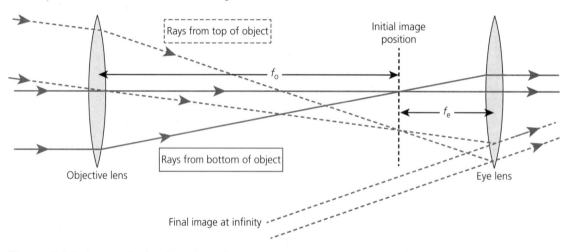

Figure 9.1 Astronomical refracting telescope

The magnification of a telescope in normal adjustment can be worked out from the formula:

$$\text{angular magnification} = \frac{\text{focal length of the objective } (f_o)}{\text{focal length of the eyepiece lens } (f_e)}$$

Angular magnification

Angular magnification is related to the increase in the angular size of an image compared with that of the object from which it was produced.

$$\text{angular magnification} = \frac{\text{angle subtended by image at the eye } (\beta)}{\text{angle subtended by the object at an unaided eye } (\alpha)}$$

(See Figure 9.2.)

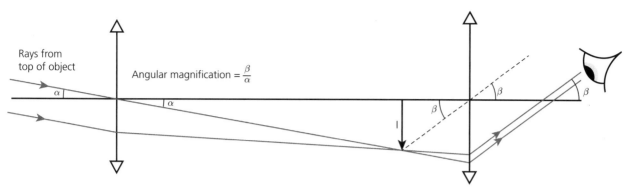

Figure 9.2 Angular magnification of a telescope

Calculate the magnification of a telescope with an objective of focal length 1200 mm using two different eyepieces of focal length:
(a) 25 mm
(b) 10 mm

Answer

(a) magnification = $\dfrac{1200}{25}$ = 48×   (b) magnification = $\dfrac{1200}{10}$ = 120×

Warning: Never look directly at the Sun, especially when using a telescope or binoculars. Blindness is likely to result.

Now test yourself

TESTED

1 You have a telescope with an objective lens of 150 cm focal length. You wish to use an eyepiece to give a magnification of 1000×. What focal length of eye lens should you use?
2 The Moon has a diameter of about 3500 km and is about 400 000 km from the Earth. What is the angle in radians that the Moon subtends to an observer on the Earth? What is this in degrees?

Answer on p. 222

Exam tip

The objective lens of a telescope has a long focal length while the eyepiece lens has a short focal length.

Reflecting telescope

REVISED

All mirrors used in astronomical telescopes are silvered on the front surface, otherwise the light would pass through the glass and result in colour distortion and multiple images. The metal used is actually aluminium, vaporised onto the surface of the glass in a vacuum. It does not reflect quite so well as silver but is better over the complete range of wavelengths of visible light.

The curvature of the mirrors is usually accurate to within one eighth of a wavelength of green light.

AQA A-level Physics   193

Cassegrain reflector

The Cassegrain reflecting telescope (Figure 9.3) has an eyepiece below the main mirror, which means that much heavier detection equipment can be fitted here. The primary (objective) mirror is large and concave; the secondary mirror is small and convex. The secondary mirror does not affect the quality of the image — the small amount of light that it interrupts is negligible compared with the total amount received by the main mirror.

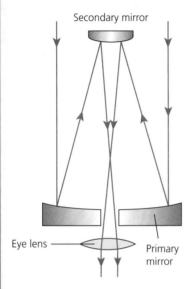

Figure 9.3 Cassegrain reflector

Comparison between refractors and reflectors

Disadvantages of a refracting telescope

- The lenses are made of glass and because the light has to go through them the glass must be perfect with no internal air bubbles.
- The lenses can only be supported around their edges but this is where they are thinnest and weakest.
- Lenses suffer from colour distortion — this means that when white light passes through them it is split into the colours of the spectrum. Because violet light refracts more than red light it is brought to a focus closer to the lens than the red light, which makes the image coloured and blurred. This effect is called **chromatic aberration** (Figure 9.4).

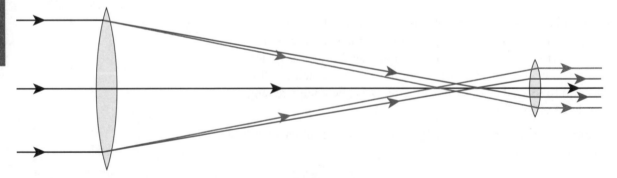

Figure 9.4 Chromatic aberration (much exaggerated)

Disadvantages of a reflecting telescope

● Rays close to the axis of the mirror are brought to a focus at one point but those far from the axis meet at a range of different points. The effective focal length varies for rays at different distances from the axis (Figure 9.5). This is called **spherical aberration** and the locus of the focal points is known as a caustic curve. This defect means that large, spherical mirrors are not good for giving a focused image over a wide field of view.

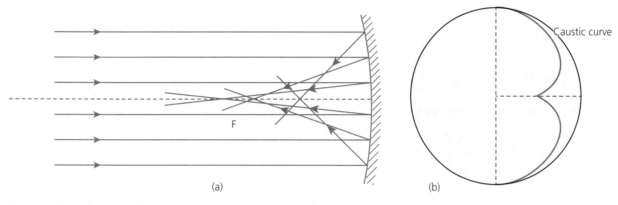

Figure 9.5 Spherical aberration (much exaggerated)

Advantages of a reflecting telescope

● The glass of the mirror does not have to be perfect throughout — only to have a perfect surface.
● The mirror can be supported across the whole of its back.
● It does not suffer from the colour defects of chromatic aberration.

For these reasons all the really large telescopes in the world today are reflectors. The largest ones have mirrors up to 10 m in diameter.

Sites for an observatory

Land-based telescopes are best placed in observatories on the tops of mountains, for the following reasons:
● They are above dust and other types of atmospheric pollution.
● They are above low cloud, mist and fog.
● They are far from light pollution from large centres of population.
● The air is thinner and so there is less atmospheric absorption.
● There are fewer convection currents in the air, so the image does not suffer so much from image shake.

The Hubble Space Telescope is in an even better position — there is no atmospheric absorption at all in space.

Telescopes using other wavelengths

REVISED

Single-dish radio telescopes

Radio telescopes operate at radio wavelengths that are considerably longer than optical wavelengths. For example, interstellar hydrogen both in our galaxy and other galaxies emits at a wavelength of 21 cm and radio telescopes are designed to detect this.

This longer operating wavelength has a number of implications for radio telescopes:

● At 21 cm the Earth's atmosphere absorbs a different proportion of the radiation.
● Radio telescopes can be used in the daytime.
● To get good resolving power a radio telescope must be very large. To achieve the same resolving power as its optical counterpart a radio telescope would have to be 50 000 times larger (see below).
● The surface of a radio telescope does not need to be so 'smooth' as that of an optical mirror. Accuracies of ±0.005 m are quite acceptable, compared with ±0.000 000 1 m for the optical mirror surface.

Exam tip

Radio telescopes have a much lower resolving power than those using visible light.

Large-diameter objectives and CCD devices

REVISED

Resolving power of telescopes

All telescopes give images that are affected by diffraction at the objective lens or mirror. This is especially important in astronomy where the images of two stars that are apparently very close together need to be separated (Figure 9.6). The aperture of the telescope needs to be as large as possible to give as little diffraction as possible and so give the telescope a high resolving power.

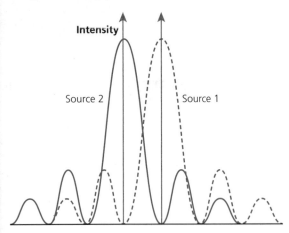

Figure 9.6 Resolution of two images in a telescope

Rayleigh criterion

This states that for two images of equal intensity to be resolved the central maximum of one diffraction pattern must fall no closer than the first minimum to the centre of the second diffraction pattern (Figure 9.6).

For a circular aperture and a small angle, the smallest resolvable angle θ (in radians) using a telescope of aperture (diameter) d operating at a wavelength λ is:

$$\theta = \frac{1.22\lambda}{d} \approx \frac{\lambda}{d}$$

Higher resolution is therefore possible with large apertures and with short-wavelength radiation such as ultraviolet light or X-rays.

The resolving power of a lens or mirror is $1.22\lambda/d$ where λ is the wavelength used and d is the aperture.

What aperture would be needed to give a resolving power of 10^{-3} radians (about 3 minutes of arc) using:
(a) a radio telescope operating at 20 cm wavelength
(b) an optical reflector operating at 600 nm wavelength?

Answer

(a) radio telescope aperture $= \dfrac{1.22 \times 0.2}{10^{-3}} = 244\,\text{m}$

(b) optical telescope aperture $= \dfrac{1.22 \times 600 \times 10^{-9}}{10^{-3}} = 0.00073\,\text{m} = 0.73\,\text{mm}$

The largest radio telescope in the world is the Arecibo telescope in Puerto Rico. It is 300 m across and is built between some small hills that form a roughly parabolic valley. The valley floor is paved with almost 40 000 aluminium panels, which act as the mirror.

Exam tip

Do not forget to use SI units in all calculations.

Now test yourself

TESTED

3 (a) What is maximum resolving power of an amateur astronomer's 15 cm reflector in the visible range (400–700 nm)?
  (b) Why will the actual resolving power be less when used to 'separate' the images of the components of a double star?

Answer on p. 223

The charge-coupled device (CCD)

A charge-coupled device (CCD) as used in astronomy is a sensitive semiconductor device that replaces traditional photographic film for recording images of astronomical objects.

The imaging area, or chip, in the CCD is composed of 'cells' made of thin layers of silicon. As light falls on a cell's surface a charge is built up. The charge depends on the total energy of the light that falls on the cell during the exposure. Therefore the longer the exposure and the brighter the light the more charge is built up on the individual cell. The cells are scanned electronically and a digital version of the image is built up. This can be downloaded to a computer for processing by the appropriate software.

The imaging area of a CCD camera consists of a grid of a large number of these cells or pixels. Cameras with over 100 million pixels on the chip are extensively used by astronomers to take long-exposure images. The chip is typically about 10 mm × 5 mm and so each pixel has an area of roughly 2.5×10^{-5} mm.

A CCD has many advantages:
- The chips can be used many times.
- They are becoming relatively cheap.
- Unwanted exposures can be discarded.
- Each image taken can be viewed more or less immediately.
- CCD devices can also be made sensitive to radiation outside the visible spectrum.

The greater number of pixels per unit area, the greater the resolution of the CCD.

Quantum efficiency of the eye and a CCD

This is defined as:

$$\text{quantum efficiency} = \frac{\text{number of electrons detected per second}}{\text{number of photons incident per second}} \times 100\%$$

A device with a high quantum efficiency will produce a large number of electrons for a given number of incident photons.

Typical values are:
- eye   1–4
- film   4–10
- CCD 70–90

This means that a CCD hugely outperforms the eye.

Classification of stars

Classification by luminosity

REVISED

Apparent magnitude (m)

The magnitude of a star is a measure of its brightness, and an approximate scale for stellar magnitudes was devised by the Greek astronomer Hipparchus. Comparing the brightness of two stars he decided that if one star was 2.5 times brighter than the other the difference of magnitude between them was 1. The modem figure is 2.51, such that a difference in brightness of 100 gives a difference in magnitude of 5 ($2.51^5 = 100$).

Bright objects have a negative apparent magnitude while faint objects have a positive apparent magnitude. The magnitude scale runs from about −26 for the Sun to around +15 for the faintest stars visible using a large astronomical telescope.

The dimmest stars that we can see with the unaided eye have an apparent magnitude of about +6.

Example

If the apparent magnitudes of two stars differ by 4 what is the difference in their brightness?

Answer

difference in brightness = $2.51^4 = 40$

Now test yourself

TESTED

4 Two stars differ in brightness by a factor of 1000. What is the difference in their apparent magnitudes?

Answer on p. 223

Luminosity and brightness

Luminosity is a measure of the total energy given output by a star at all wavelengths, from gamma radiation to radio waves. This luminosity depends on:

- the size of the star
- the temperature of the star

The brightness is how bright a star appears when seen from the Earth. This depends on:

- the actual luminosity of the star
- the distance of the star from the observer on the Earth

Distance measurement and absolute magnitude

Astronomical unit, parsec and light year

1 **astronomical unit** (AU) is the mean distance of the Earth from the Sun.

1 **parsec** is the distance at which an object subtends an angle of 1 arc second using the radius of the Earth's orbit as the baseline (Figure 9.7).

1 light year is the distance that light travels in free space in 1 year.

1 light year (= 9.46×10^{15} m)

1 parsec = 3.086×10^{16} m = 3.26 light years

1 megaparsec (Mpc) = 3.26×10^6 light years = 3.086×10^{22} m

Absolute magnitude (M)

How bright a star looks is given by its apparent magnitude. This is different from its **absolute magnitude**. The apparent magnitude (m), the distance of the star in parsecs (d) and the absolute magnitude (M) are related by the equation:

$$m - M = 5\log\frac{d}{10} \text{ or } M = m - 5\log\frac{d}{10}$$

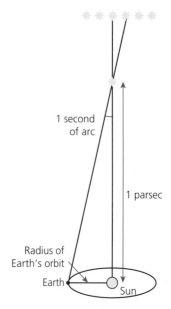

Figure 9.7 Definition of the parsec

> The **absolute magnitude** of a star is defined as the apparent magnitude that it would have if placed at a distance of 10 parsecs from the Earth.

> **Exam tip**
>
> The greater the luminosity of a star, the lower its magnitude value (including negative).

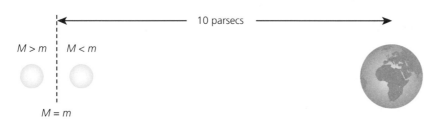

Figure 9.8 Apparent magnitude (m) and absolute magnitude (M)

Table 9.1 gives some examples of the magnitudes and distances of five bright stars.

Table 9.1

Star name	Distance/pc	Apparent magnitude	Absolute magnitude
Sirius	2.7	−1.46	1.4
Vega	8.1	0.03	0.6
Arcturus	11	−0.04	−0.2
Aldebaran	21	0.85	−0.3
Rigel*	260	0.12	−7.1

*Rigel is a variable star and so its absolute magnitude and therefore also its apparent magnitude change slightly with time.

Example 1

Calculate the absolute magnitude of a star of apparent magnitude +2.5 that is at a distance of 25 pc from the Earth.

Answer

$$M = m - 5\log\frac{d}{10} = 2.5 - 5\log 2.5 = 2.5 - 1.99 = +0.51$$

Exam tip

Distances (d and D) should be in parsecs.

Example 2

Calculate the distance of a star with an apparent magnitude of +6.0 and an absolute magnitude of +4.0.

Answer

$$6 - 4 = 5\log\frac{d}{10}$$

$$\frac{2}{5} = \log\frac{d}{10} = 0.4$$

$$\frac{d}{10} = 2.5$$

$$d = 25\,\text{pc}$$

Now test yourself

TESTED

5 (a) A star is 20 pc from the Earth and has an apparent magnitude of −0.7. What is the absolute magnitude of the star?
  (b) A star is 5 pc from the Earth and has an absolute magnitude of +2.6. What is the apparent magnitude of the star?
  (c) A star has an apparent magnitude of −0.2 and an absolute magnitude of +2.0. What is the distance of the star from Earth in light years? (1 light year = 0.31 parsecs)

Answer on p. 223

Classification by temperature

REVISED

Black body radiation

The amount of infrared radiation emitted by a body depends on three things:
- the surface area of the body
- the type of surface
- the temperature of the body

Simple experiments show that rough, black surfaces make the best emitters and absorbers of radiation at a given temperature.

An ideal absorber would be one that absorbed *all* the radiation that fell on it, and also one that emitted the maximum amount of radiation possible for that area at that temperature. Such a body is known as a black body, and the radiation emitted by it as **black body radiation**.

Wien's displacement law

Figure 9.9 shows the variation of energy emitted with wavelength for two stars at different temperatures (assumed to behave like black bodies).

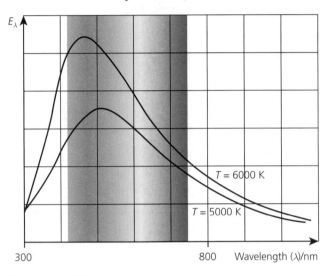

Figure 9.9 Wien's displacement law

Some important facts can be deduced from these curves:
- The area between any energy–wavelength curve and the wavelength axis is proportional to the total energy emitted by the body per unit area at that temperature.
- The maxima of the curves move towards shorter wavelengths at higher temperatures.
- The curves for lower temperatures lie completely inside those of higher temperature.

$$\lambda_m T = \text{constant} = 2.90 \times 10^{-3}\,\text{m K}$$

where λ_m is the wavelength at which most energy is emitted, that is, at the maximum of the curve.

Stefan's law

Stefan's law relates the luminosity of a star to its surface area (A) and surface temperature (T):

$$\text{power (luminosity)} = \sigma A T^4$$

σ is Stefan's constant and has a value of $5.67 \times 10^{-8}\,\text{W m}^{-2}\,\text{K}^{-4}$.

> **Example 1**
>
> Calculate the luminosity of a star of radius $10^9\,\text{m}$ (slightly larger than our Sun) and surface temperature $6000\,\text{K}$.
>
> **Answer**
>
> luminosity $= \sigma A T^4 = 5.67 \times 10^{-8} \times 4 \times \pi \times 10^{18} \times 6000^4 = 9.23 \times 10^{26}\,\text{W}$

> **Example 2**
>
> Calculate the temperature of a star with a maximum energy emission at a wavelength of 400 nm.
>
> Answer
>
> temperature $(T) = \dfrac{2.90 \times 10^{-3}}{400 \times 10^{-9}} = 7250\,K$

Now test yourself

TESTED

6 Calculate the temperature of a star that has a luminosity of 10^{30} W and a radius of 10^{11} m. (Stefan's constant = $5.67 \times 10^{-8}\,W\,m^{-2}\,K^{-4}$)

Answer on p. 223

Use of stellar spectral classes

REVISED

By looking at the spectrum of a star astronomers can determine:
- the temperature of the star. This can be found by measuring the variation in intensity across the spectrum. When the wavelength of the peak intensity is found the temperature can be calculated using Wien's law.
- the composition of the star. The chemical composition of the star can be determined by looking at the absorption lines in the spectrum. These indicate the presence of particular elements in the star.

The spectra of stars are classified into a number of types, with each type of star having a letter (Table 9.2).

Table 9.2

Spectral class	Intrinsic colour	Temperature/K	Prominent absorption lines
O	Blue	25000–50000	He$^+$, He, H
B	Blue	11000–25000	He, H
A	Blue–white	7500–11000	H (strongest), ionised metals
F	White	6000–7500	ionised metals
G	Yellow–white	5000–6000	ionised and neutral metals
K	Orange	3500–5000	neutral metals
M	Red	<3500	neutral atoms, TiO

This list ranges from very hot O-type stars to 'cool' M-type stars. O and B stars are blue–white and M stars reddish in appearance when viewed from the Earth.

The temperature of the star can be found by measuring the Doppler broadening of the spectral lines (see page 213).

The Hertzsprung–Russell (HR) diagram

General shape of the HR diagram

This type of diagram originally showed the variation of the absolute magnitude of a star against its spectral type.

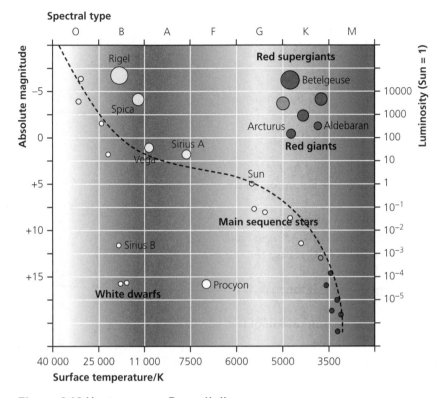

Figure 9.10 Hertzsprung–Russell diagram

Hot, bright stars are towards the top-left of the diagram while cool and dim stars are towards the bottom-right. Some large stars, like the red supergiants, are quite cool, but because of their enormous size they are in the top-right of the diagram. On the other hand, white dwarfs are hot, small stars and so appear to the bottom-left.

There is a band of stars that runs from the top-left to the bottom-right (shown by the dotted line in the diagram). This is called the **main sequence** and stars that lie in this area are called main-sequence stars. These are 'normal' stars, while those that lie to one side or other of this area are 'unusual' stars and include white dwarfs, red giants and supergiants. Supergiant stars can be either hot (e.g. Rigel) or cool (e.g. Betelgeuse). About 90% of the stars in our region of the galaxy are main-sequence stars (including our Sun); 10% are white dwarfs and 1% are red giants or supergiants.

Evolution of stars of similar mass to our Sun (< about 3.0 M$_\odot$)

The life cycle of a star with a mass of less than 3.0 solar masses (M$_\odot$) can be summarised as follows:

1 Particles begin to clump together in a low-density gas cloud.
2 Nuclear fusion begins and the star is 'born'.
3 Fusion of light elements takes place and the star reaches a stable state in its 'main sequence phase'.

4 Fusion of hydrogen in the core eventually ceases but continues in the outer shell. This gives an outward pressure so that the star expands to form a red giant with a high luminosity but lower surface temperature.
5 Eventually layers of the star are blown away as a planetary nebula, leaving a core.
6 The star core radiates less and shrinks to become a white dwarf and finally a cold black dwarf.

Evolution of stars of mass many times that of our Sun (> about 3 M$_\odot$)

The life cycle of a star with a mass of more than 3 M$_\odot$ can be summarised as follows:
1 Particles begin to clump together in a low-density gas cloud.
2 Nuclear fusion begins and the star is 'born'.
3 Fusion of heavier elements takes place — up to iron and nickel.
4 As with lighter stars, the star then passes through a red giant or, in this case, a red supergiant phase.
5 Eventually the gravitational attraction exceeds the radiation pressure and a catastrophic collapse occurs. The outer layers collapse in seconds, reaching speeds of up to 30% of the speed of light. In this case the shock-wave moves outwards, resulting in a supernova.
6 A neutron star is formed, which then changes further if the mass is greater than about ten solar masses.
7 For a star of 10 or more solar masses, gravitational compression of the neutron star continues, eventually forming a black hole.

The evolutions of stars is outlined in Figure 9.11.

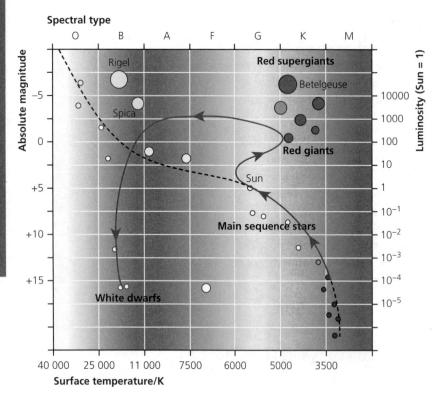

Figure 9.11 Evolution of stars

Supernovae, neutron stars and black holes

Supernovae

A supernova is the explosive cataclysmic end to the life of a star. A supernova can produce so much energy that it can briefly outshine a galaxy with about 10^{10} times the luminosity of the Sun! The energy produced in a supernova explosion can be around 10^{46} J. This enormous release of energy may occur in only a few seconds.

A star of mass greater than 1.4 $M_\odot$ can become a supernova, but the precise value depends on how it was formed).

Type 1a (1.4 $M_\odot < m < 3$ $M_\odot$)

A type 1a supernova results from the attraction of material to a white dwarf from its companion star in a binary system, increasing both its mass and core density.

If the mass of the white dwarf reaches more than about 1.4 $M_\odot$ with a correspondingly 'super-high' density the star then suffers a violent collapse, with a huge rise in temperature. The thermal energy produced causes the star to explode as a supernova; no individual body remains.

Type 1a supernovae are used as **standard candles** to determine galactic distances. Since they all have approximately the same absolute magnitude, a measurement of their apparent magnitude can be used to calculate their distance.

The light curve for a typical type 1a supernova is shown in Figure 9.12.

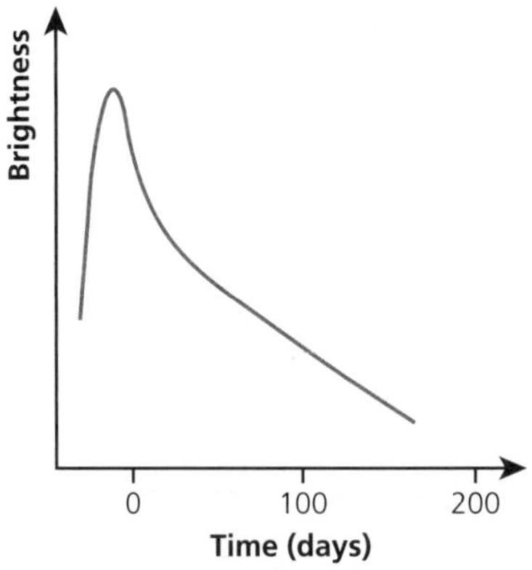

Figure 9.12 Light curve of a type 1a supernova

Type II (8 $M_\odot < m < 50$ $M_\odot$)

A type II supernova may result in the latter stages of the life of a single giant star. In the type II supernova the gravitational attraction overcomes the radiation pressure as the star 'runs out of fuel'. The outer layers of the star fall inwards, bounce off the solid core and travel outwards towards the surface. The shock wave produced moves at up to $10\,000\,\mathrm{km\,s^{-1}}$ and 'blows away' most of the outer layers of the star, giving a supernova explosion. If the final mass of the core is less than about 1.4 $M_\odot$ a neutron star is left behind. If it is greater it collapses to give a black hole.

During the supernova phase, as a rapidly rotating high mass star collapses to form a neutron star, very high intensity gamma ray emissions occur. These are known as gamma ray bursts (GRBs). A GRB may last for several hours and produce as much energy in a few seconds as the Sun will emit in 10 billion years.

Neutron stars (3 $M_\odot < m < 10$ $M_\odot$)

A neutron star is the final phase of a star with a mass of between 3 $M_\odot$ and 10 $M_\odot$ (where $M_\odot$ is the mass of the Sun). The star contracts, the density increasing to the same as that of an atomic nucleus and electrons combine with protons to give matter composed only of neutrons. Neutron stars have a very high magnetic field and a surface gravity up to 10^{11} that of Earth.

Black holes ($m > 10$ $M_\odot$)

A black hole is an unimaginable concentration of matter formed by the gravitational collapse of a massive star, usually greater than 10 solar masses. On approaching a black hole the escape velocity increases and increases. Eventually a point will be reached where the escape velocity is the speed of light — this point is known as the event horizon for the black hole. The radius of the sphere of the **event horizon** around the black hole is known as the **Schwarzschild radius** (R_S).

At the Schwarzschild radius, the escape velocity (V_E) is given by:

$$v_E = c = \sqrt{\frac{2GM}{R_S}}$$

where c the speed of light, G the gravitational constant and M the mass of the star.

At the event horizon, a distance of R_S from the centre of the black hole, the escape velocity will be equal to the speed of light.

Example

Calculate the radius of the event horizon for a black hole with a mass 50 times that of our Sun. ($G = 6.67 \times 10^{-11} \, \text{N} \, \text{m}^2 \, \text{kg}^{-2}$; mass of the Sun $= 2 \times 10^{30} \, \text{kg}$; speed of light $= 3 \times 10^8 \, \text{m} \, \text{s}^{-1}$)

Answer

$$R_s = \frac{2GM}{c^2} = \frac{2 \times 6.67 \times 10^{-11} \times 50 \times 2 \times 10^{30}}{9 \times 10^{16}} = 1.5 \times 10^5 \, \text{m} = 150 \, \text{km}$$

Black holes swallow up matter — increasing in mass themselves — and so their Schwarzschild radius increases as the event horizon expands. However, if antimatter falls into a black hole then the mass of the black hole decreases and eventually disappears in a burst of radiation.

Figure 9.13 represents the gravitational fields outside a large star and a black hole. The depth of the gravitational vortex is much greater for the black hole than for a heavy star — the escape velocity at the event horizon being the speed of light.

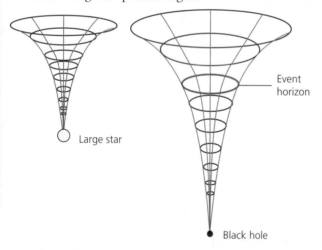

Figure 9.13 Gravitational potential curves for a large star and a black hole

Revision activity

Draw a mind map to show the stages in the formation of a black hole.

Now test yourself

TESTED

7 Calculate the mass of a black hole that has a Schwarzschild radius of 200 km. ($G = 6.67 \times 10^{-11} \, \text{N} \, \text{m}^2 \, \text{kg}^{-2}$; speed of light $= 3 \times 10^8 \, \text{m} \, \text{s}^{-1}$)

Answer on p. 223

Cosmology

Doppler effect

The Doppler effect (Figure 9.14) is the apparent change of frequency and wavelength when a source of waves and an observer move relative to each other. The Doppler shifts in frequency (Δf) and wavelength ($\Delta \lambda$) are given by the following equations:

$$\textbf{doppler shift} = \frac{\Delta\lambda}{\lambda} = \frac{v}{c}$$

$$\textbf{doppler shift} = \frac{\Delta f}{f} = \frac{v}{c}$$

where v is the speed of the source and c is the speed of the waves.

In astrophysics astronomers define the quantity v/c as the **red shift** (Figure 9.15) and give it the symbol z.

The velocity of the star along a line of sight joining the star to the Earth can be determined by comparing the spectral lines due to a certain element with the spectral lines of the same element produced in the laboratory. The shift of the lines can be measured. Knowing the shift of the lines the velocity of the star can be found using the equations for the Doppler effect.

In astrophysics the Doppler effect can also be used to find:
- the speed of one component of a double star system around the other
- the recession or approach speed of a galaxy or a quasar
- the temperature of the star by the broadening of the absorption lines in the spectrum

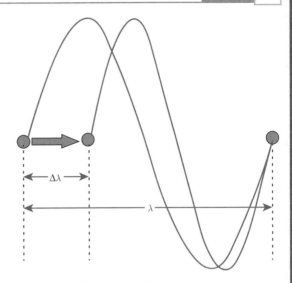

Figure 9.14 Doppler effect

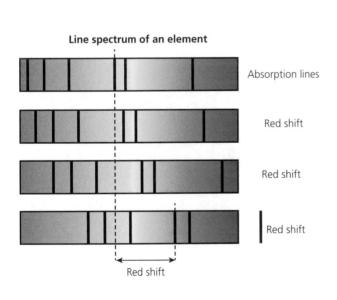

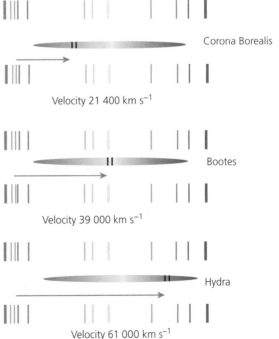

Figure 9.15 Galactic red shift

Observations of the spectra of galaxies show that the light coming from many of these is shifted significantly towards the red, which shows that they are moving away from us at high speeds — many tens of thousands of kilometres per second. If the Doppler shift of lines within their spectra can be measured, their speed of recession can be calculated.

For very high speeds the simple formula cannot be used and the effects of special relativity have to be allowed for.

> **Exam tip**
>
> Notice that the wavelength and frequency shifts depend on the original wavelength or frequency — red light is shifted more than blue for a given velocity — and that these changes apply whether the source or observer, or both, are moving.

Example 1

A galaxy is moving away from the Earth at 26 000 km s^{-1}. Calculate the wavelength and frequency change of a 650 nm line in its spectrum. Take $c = 3 \times 10^8$ m s^{-1}.

Answer

wavelength change $(\Delta\lambda) = \dfrac{\lambda v}{c} = \dfrac{650 \times 26\,000 \times 10^3}{3 \times 10^8} = 56.3$ nm

frequency change $(\Delta f) = \dfrac{f v}{c} = \dfrac{4.6 \times 10^{14} \times 26\,000 \times 10^3}{3 \times 10^8} = 0.4 \times 10^{14}$ Hz

Example 2

A galaxy is moving relative to the Earth and a 600 nm line in its spectrum is found to be Doppler shifted by 30 nm towards the red.
(a) Is the galaxy receding or approaching?
(b) Calculate its speed relative to the Earth.

Answer

(a) Receding from the Earth, as shown by the shift towards the longer wavelength — red.

(b) speed of recession $= \dfrac{c\Delta\lambda}{\lambda} = \dfrac{3 \times 10^8 \times 30 \times 10^{-9}}{600 \times 10^{-9}} = 1.5 \times 10^7$ m s^{-1}

The Doppler effect can also be used to measure the speed of one component of a binary star system about another, as shown in Figure 9.16.

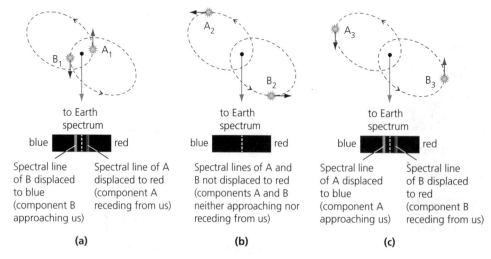

Figure 9.16 A binary star system — eclipsing variable

Exam practice answers and quick quizzes at **www.hoddereducation.co.uk/myrevisionnotes**

Now test yourself

8 IC3258 is a galaxy in the Virgo cluster that is moving towards the Earth at $500\,km\,s^{-1}$. A line in its spectrum is measured to have a wavelength of 600 nm.
   (a) Has this line been red shifted or blue shifted?
   (b) What wavelength would be measured if the galaxy was at rest relative to the Earth?
   (c) What is the Doppler shift of this line?

Answer on p. 223

Hubble's law

The Hubble formula provides a very powerful way of determining not only distances of remote galaxies but also the age of the universe itself.

$$H_0 = \frac{v}{d} \qquad v = H_0 d$$

where H_0 is the Hubble constant, v is the recession speed of the galaxy in $km\,s^{-1}$ and d is the distance of the galaxy from Earth in Mpc. The value of the Hubble constant will vary with time and measurement technique but at present:

$$H_0 \approx 71.5\,km\,s^{-1}\,Mpc^{-1}$$

This means that the velocity of recession of a galaxy increases by $71.5\,km\,s^{-1}$ for every 1 Mpc increase in its distance.

Using Hubble's law the red shift (z) can be written as:

$$z = \frac{v}{c} = H_0 \frac{d}{c}$$

Conversion of Hubble constant to SI units

Take the Hubble constant H_0 to be $71.5\,km\,s^{-1}\,Mpc^{-1}$ and 1 light year to be $9.46 \times 10^{15}\,m$:

$$1\text{ parsec} = 3.26\text{ light years} = 3.086 \times 10^{16}\,m$$

Therefore: $1\text{ Mpc} = 3.086 \times 10^{22}\,m$

So:

$$71.5\,km\,s^{-1}\,Mpc^{-1} = \frac{71.5 \times 10^3}{3.086 \times 10^{16} \times 10^6} = 2.32 \times 10^{-18}\,s^{-1}$$

Example

Find the distance r of the galaxy with a recession velocity of $1200\,km\,s^{-1}$ if the Hubble constant is $71.5\,km\,s^{-1}\,Mpc^{-1}$.

Answer

Using $v = H_0 r$:

$$r = \frac{v}{H_0} = \frac{1200}{71.5} = 16.78\,Mpc = 16.78 \times 3.086 \times 10^{22} = 5.19 \times 10^{23}\,m$$

$$= 5.5 \times 10^7\text{ light years}$$

The galaxy is therefore 55 million light years away.

Exam tip

It is important to realise that the number quoted above as the value of H_0 is the value at the present time. The value of H_0 will have varied over the lifetime of the universe and will continue to do so in the future.

9 The maximum speed of a galaxy relative to the Earth is the speed of light ($3 \times 10^8\,\mathrm{m\,s^{-1}}$). Use Hubble's law to calculate the maximum radius of the observable universe. Give your answer in both metres and light years.

Answer on p. 223

The expansion and age of the universe

If we assume that:
- the radius of the universe (R) = velocity of recession of the most distant galaxies (v) × the age of the universe (t_0)
- $v = H_0 R$
- the galaxies have been moving apart with constant velocity since the beginning of time

the age of the universe (t_0) can be found from the equation:

$$\text{age of the universe } (t_0) = \frac{R}{v} = \frac{1}{H_0}$$

$$t_0 = \left(\frac{1}{71.5 \times 10^3}\right) \times 3.086 \times 10^{22} = 4.32 \times 10^{17}\,\mathrm{s} = 1.37 \times 10^{10} \text{ years}$$

$$= 13.7 \text{ thousand million years}$$

Since $t_0 = 1/H_0$, a large value of H_0 implies a young universe.

The Big Bang

It is now generally accepted by most astronomers that the universe as we know it began with an unimaginably huge expansion from an unimaginably small, unimaginably hot and dense point, between 13.7 billion (1.37×10^{10}) and 13.8 billion years ago. We call this the Big Bang.

Time and space both originated at the same time, with the Big Bang. Before that there was no space and no time — the Big Bang 'created' space and time. We cannot ask what happened before the Big Bang because before that moment nothing existed — no space and no time!

What happened afterwards is outlined in Table 9.3.

Table 9.3

Time after the Big Bang	Nature of the universe	Temperature
10^{-43} s		
10^{-34}–10^{10} s		
10^{-10} s	Particle soup dominates	10^{15} K
1 s	Neutrons and protons formed	10^{10} K
3 minutes	Helium nuclei formed	10^9 K
300 000 years	Microwave background fills the universe	6000 K
500 000 years	Temperature falls further; infrared	750 K
1 million years	Atoms form; stars and galaxies exist; universe becomes transparent	
1 billion (10^9) years	The first stars; heavy elements form	18 K (−255°C)
14 billion years	The present day	2.7 K (−270.3°C)

Cosmic microwave background radiation

In 1965 two American astronomers, Arno Penzias and Robert Wilson, were using the antenna at the Bell Laboratories in New Jersey for scanning the sky when they found that there was a background 'noise' (like static in a radio). This uniform signal was in the microwave range, with a peak at a wavelength of about 2 mm.

By pointing the 'telescope' in a variety of directions they concluded that the interference was not radiation from our galaxy or from extra-terrestrial radio sources, and because it remained constant throughout the year it could not have come from the solar system. It seemed to come from all parts of the sky.

Finally they realised that it was not random noise causing the signal but something that pervaded the whole universe. This was **cosmic microwave background** (CMB) radiation with a 'temperature' of around 2.7 K, and was given an evocative name — the 'echo of the Big Bang'. It is the residual radiation predicted by Gamov and others, and is the result of the universe cooling from an unimaginably hot state over the intervening 13 billion years. The detection of CMB radiation supports the Big Bang idea of the universe, because the cooling of the universe after the Big Bang would suggest an expansion over many millions of years.

Exam tip

The discovery of small discontinuities in the CMB is very good evidence for the birth of galaxies.

Critical density of the universe

The Hubble constant (H_0) is also important in predicting the ultimate fate of our universe, which relates to the concept of its critical density. The velocity of recession of a galaxy can be considered as its escape velocity from the rest of the universe:

velocity of recession (v) = $H_0 R$

where R is the distance of the galaxy.

However:

escape velocity (v) = $\sqrt{\dfrac{2GM}{R}}$

Therefore:

$$\rho = \frac{3H_0^2}{8\pi G}$$

where ρ is the critical density of the universe and G is the gravitational constant.

(This simplified calculation assumes both constant v and a constant value of the Hubble constant (H_0).)

If the density of the universe is greater than this the universe will contract; if it is less, it will expand for ever.

> **Example**
>
> Calculate the critical density of the universe. ($H_0 = 71.5\,\text{km}\,\text{s}^{-1}\,\text{Mpc}^{-1}$; $G = 6.67 \times 10^{-11}\,\text{N}\,\text{m}^2\,\text{kg}^{-2}$)
>
> **Answer**
>
> $$\text{critical density } (\rho) = \frac{3H_0^2}{8\pi G} = \frac{(2.32 \times 10^{-18})^2 \times 3}{8 \times 3.14 \times 6.67 \times 10^{-11}} = 9.63 \times 10^{-27}\,\text{kg}\,\text{m}^{-3}$$
>
> The mass of a proton is $1.66 \times 10^{-27}\,\text{kg}$, so this density is equivalent to almost six (5.8) protons in every cubic metre of space.

Now test yourself

TESTED

10 If the value of the Hubble constant were $70\,\text{km}\,\text{s}^{-1}\,\text{Mpc}^{-1}$ (a little lower than the currently accepted value) calculate the critical density of the universe.

Answer on p. 223

The fate of the Universe

At the time of writing (2017) the future state of the universe is uncertain. It is thought that the universe is expanding and that this expansion is accelerating (shown by the blue line in Figure 9.17). This expansion appears to be affected by a form of energy known as **dark energy**. We are not sure what this is, how long it has been active or whether the effects of it are constant or changing as time passes.

We define a quantity Ω where:

$$\Omega = \frac{\text{actual average density of matter in the universe}}{\text{critical density of the universe}}$$

The critical density will decide whether the universe is:

- open — a runaway expansion ($\Omega < 1$)
- flat — an expansion but slowing ($\Omega = 1$)
- closed — a final contraction known as the Big Crunch ($\Omega > 1$)

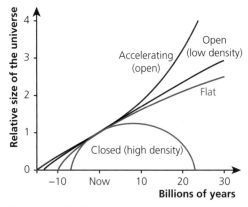

Figure 9.17 The fate of the universe

Quasars

REVISED

Quasars (Figure 9.18) are exceptionally luminous star-like sources of radiation, with very high red shifts.

They have intensities between 10^{12} and 10^{15} times that of the Sun, which means that a single quasar may emit about 1000 times as much energy as our whole galaxy. The quasar 3C 273 has an absolute magnitude of -27. It is because of this enormous rate of emission of energy that the lifetimes of quasars are relatively short.

The radiation from quasars is thought to originate from a vast disc of gas known as an accretion disc, which has been attracted towards the event horizon of a supermassive black hole at the centre of a galaxy. The enormous gravitational pull of the black hole causes the gas in the disc to rotate at high speed, finally spiralling into the black hole.

The enormous speeds of molecules within this gas cloud generate a fantastically high temperature, which results in the emission of light of unimaginable intensity.

> **Exam tip**
>
> $z = v/c$ cannot be used for quasars because the very high value of v means relativistic effects.

> **Exam tip**
>
> Quasars emit all wavelengths of the electromagnetic spectrum strongly, from gamma rays to radio waves but especially in the radio frequency range.

Figure 9.18 A quasar

Detection of exoplanets

REVISED

An exoplanet (Figure 9.19) is a planet orbiting a star other than our Sun. At the time of writing (2016) exoplanets have only been detected within our own galaxy.

Methods of detection include:

- Variation in the Doppler shift of the star as the exoplanet and star orbit a common centre of gravity. Only observable for very large planets.
- Transit of the planet across the face of the star, causing a slight dip in the intensity of the light from the star received by the observer.
- Changes in the observed spectrum of the star due to absorption of the starlight by the atmosphere of the exoplanet during transit.
- Direct observation by telescopic imaging.

There are difficulties in detecting exoplanets:
- It involves observation of relatively small objects at a great distance (at least 10 light years).
- The brightness of the star about which the planet is orbiting.

On 23 July 2015 the discovery of an exoplanet in orbit round a star in the constellation Cygnus was announced. Named Kepler 452B, it was detected because the light from its parent star shows a slight dip when the planet passes in front of it. Kepler 452B seems to be very similar to our Earth; the surface temperature suggests that water could exist there as a liquid, and the length of the planet's year is 385 Earth days. This planet is a billion years older than the Earth, which may mean that life could already have developed on it, and is around 1400 light years (430 pc) away from us.

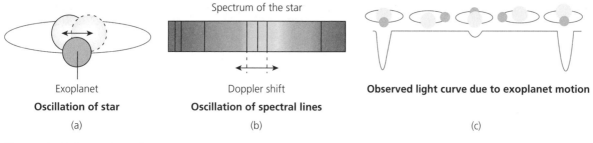

Figure 9.19 Detection of an exoplanet

AQA A-level Physics   213

9 Astrophysics

Now test yourself

11 Estimate the change in intensity of light received from a star–exoplanet system when the planet passes in front of the star if the radius of the planet is 10% of the radius of the star.

Answer on p. 223

Exam practice

1 A simple astronomical refracting telescope has two lenses, an objective lens and an eyepiece lens. Which statement best describes these two lenses?
   A The objective lens has a smaller focal length and smaller power than the eyepiece lens.
   B The objective lens has a larger focal length and larger power than the eyepiece lens.
   C The objective lens has a larger focal length and smaller power than the eyepiece lens.
   D The objective lens has a smaller focal length and smaller power than the eyepiece lens. [1]
2 Under certain conditions the human eye has an aperture with a diameter of 0.4 cm. Calculate the resolving power of the eye for a wavelength of 600 nm in this state. [2]
3 The event horizon of a black hole is:
   A the radius of the black hole
   B the radial distance from the centre of a black hole to where the escape velocity is equal to that of light
   C the radial distance from the centre of a black hole to where the escape velocity is zero
   D the initial radius of the supermassive star that collapsed to form the black hole [1]
4 Which of the following statements is correct?
   A The absolute magnitude of a star is defined as the apparent magnitude that it would have if placed at a distance of 10 light years from the Earth.
   B The apparent magnitude of a star is defined as the absolute magnitude that it would have if placed at a distance of 10 pc from the Earth.
   C A star at a distance of 15 pc from the Earth and an apparent magnitude of +5 will have an absolute magnitude that is more positive.
   D The absolute magnitude of a star is defined as the apparent magnitude that it would have if placed at a distance of 10 pc from the Earth. [1]
5 (a) Calculate the luminosity of a star of radius 10^{10} m and surface temperature 5000 K. [3]
   (b) Calculate the temperature of a star with maximum energy at a wavelength of 550 nm in its energy distribution. [2]
6 A certain spectroscopic binary has a massive central star with a smaller one moving round it in a circular orbit. The orbit period of the smaller star is 2.86 days and when it is moving in the line of sight as seen from the Earth it has a velocity towards the Earth of 50 km s^{-1}.
   (a) What will be the maximum Doppler shift of a line of wavelength 650 nm in the spectrum of the smaller star? [2]
   (b) Sketch the way in which this shift will vary over the time of one complete orbit. [2]
7 The H_α spectral line, wavelength 656 nm, observed in the spectrum of the galaxy 3C 273 is Doppler shifted towards the red by 0.1 μm. Calculate:
   (a) the speed of recession of 3C 273 [2]
   (b) the distance of 3C 273 when the Doppler shift was measured [2]
   Hubble constant = 2.31×10^{-18} s^{-1}
8 How would a higher value of the Hubble constant affect the predicted amount of dark energy in the universe? Explain your answer. [3]

Answers and quick quiz 9 online

Summary

You should now have an understanding of:

- Telescopes — magnification and focal lengths of objective and eyepiece
- Reflecting telescopes — Cassegrain reflector, comparison between refractors and reflectors
- Advantages of large-diameter telescopes — greater resolution and light-gathering power
- Classification of stars by luminosity — Wien's and Stefan's laws
- Absolute and apparent magnitude — equation relating these quantities
- Classification by temperature — black body radiation; Wien's and Stefan's laws
- Stellar spectral classes — colour, temperature and absorption lines
- Hertzsprung–Russell (HR) diagram — main sequence, dwarfs and giants, stellar evolution
- Supernovae, neutron stars and black holes
- Doppler effect — optical and radio frequencies, double stars, galaxies and quasars
- Hubble's law — red shift, big bang theory, age and size of the universe
- Quasars — bright radio sources, large red shifts, formation from active black holes
- Detection of exoplanets — methods and difficulties of detection

Now test yourself answers

Chapter 1

1. $0.000025\,J = 2.5 \times 10^{-4}\,J$
2. $\text{pressure} = \dfrac{\text{force}}{\text{area}} = \dfrac{\text{mass} \times \text{acceleration}}{\text{area}} =$
   $\text{kg}\,\text{m}\,\text{s}^{-2}/\text{m}^2 = \text{kg}\,\text{m}^{-1}\,\text{s}^{-2}$

Chapter 2

1. (a) 143
   (b) 146
2. 54
3. (a) $^{14}_{7}\text{N}$
   (b) $^{14}_{6}\text{C} \rightarrow\ ^{14}_{7}\text{N} +\ ^{0}_{-1}\text{e} + \bar{\nu}_e$
4. (a) alpha
   (b) $^{238}_{92}\text{U} \rightarrow\ ^{234}_{90}\text{Th} +\ ^{4}_{2}\text{He}$
5. (a) $2 \times 10^{-13}\,J$
   (b) $2 \times 10^{-15}\,J$
   (c) $4.74 \times 10^{-19}\,J$
   (d) $3.32 \times 10^{-19}\,J$
   (e) $2.84 \times 10^{-19}\,J$
   (f) $9.95 \times 10^{-24}\,J$
   (g) $7.83 \times 10^{-28}\,J$
6. Charge: $-1 = +1 + 0 + 0$
   Lepton number: $1 = 1 + 1 + (-1)$
7. $1.33 \times 10^{-18}\,J$
8. (a) $1.66 \times 10^{-18}\,J$
   (b) $1.0 \times 10^{-18}\,J$
   (c) $1.5 \times 10^{15}\,Hz$
9. $\text{frequency} = \dfrac{E_2 - E_1}{h} = 2.46 \times 10^{15}\,Hz$
   $\text{wavelength} = c/f = 122\,nm$
10. $\text{wavelength} = h/mv = 3.97 \times 10^{-14}\,m$
    $= 3.97 \times 10^{-5}\,nm$

Chapter 3

1. $2.94\,m$
2. Light is electromagnetic radiation (varying electric and magnetic fields) and this will travel through a vacuum. Sound requires a medium to transmit it and so would not be transmitted though the virtually airless 'atmosphere' of the Moon.
3. (a) They would be aligned with the bars parallel to each other.
   (b) The received signal would vary in intensity sinusoidally.
4. half a wavelength
5. The ends of a stretched string will be fixed points.
6. The received signal is the result of the superposition of the signals from the two transmitters.
   $\text{distance between maxima} = \dfrac{\lambda}{2} = 30 \times 5 = 150\,m$;
   $\text{wavelength} = 300\,m$
7. (a) $\text{fringe width} = \lambda D/s$
   $= \dfrac{600 \times 10^{-9} \times 0.9}{0.5 \times 10^{-3}} = 0.108\,mm$
   (b) moved further away by 18 cm
8. $\text{wavelength} = \dfrac{4.5 \times 10^{-3} \times 0.6 \times 10^{-3}}{4}$
   $= 6.75 \times 10^{-7}\,m = 675\,nm$
9. Light has a much shorter wavelength.
10. Yes — sound waves show diffraction.
11. (a) $d = \dfrac{1}{250\,000} = 4 \times 10^{-6}\,m$
    $n\lambda = d\sin\theta$
    $\sin\theta = \dfrac{550 \times 10^{-9}}{4 \times 10^{-6}} = 0.1375$
    $\theta = 7.9°$
    (b) $n = \dfrac{d\sin 90}{550 \times 10^{-9}} = 7$
    (c) Seven either side of a central image (15).
12. Spectra wider because of the finer 'grating spacing'.
13. $n_1\sin\theta_1 = n_2\sin\theta_2$
    refractive index $(n_1) = 1.47$
14. $\text{critical angle} = \dfrac{n_d}{n_w} = 33.3°$
15. critical angle for the glass = 42°
    maximum angle of incidence at one face $\leq 90°$
    maximum angle of incidence at the adjacent face in the glass = $90 - 42 = 48°$
    Therefore total internal reflection will take place at this face and so the light will not emerge.
16. Different wavelengths of light will have different refractive indices. Therefore these different wavelengths will travel at different speeds in the fibre. This will give pulse broadening.

Chapter 4

1. bearing 8.2°; speed 353 m s⁻¹

2. horizontal component = 2000 cos 15 = 1932 N
   vertical component = 2000 sin 15 = 518 N

3. Using moment = $Fd \sin \theta$:
   (a) 9.6 N m
   (b) 15 N m
   (c) 34 N m

4. $140 = 0.225 \times F$
   $F = 622$ N

5. Take moments about P:
   $(1.5 \times 6 \times 9.8) + (3 \times 8) = 112.2 = T \cos 35 \times 2.4$
   $T = 57.1$ N

6. (a) 1050 N
   (b) It will be unchanged.

7. (a) 271 m
   (b) 4.4 m s⁻¹
   (c) −0.07 m s⁻²

8. 216 s (3 m 36 s)

9. $a = \dfrac{v^2}{2s} = \dfrac{10^{14}}{2 \times 3 \times 10^{-2}} = 1.67 \times 10^{15}$ m s⁻²

10. (a) $s = \frac{1}{2}gt^2 = 78.4$ m
    (b) $v = gt = 39.2$ m s⁻¹
    (c) $v_{average} = 19.6$ m s⁻¹

11. (a) (i) 200 m s⁻¹
        (ii) 200 m s⁻¹
    (b) (i) $v = gt = 19.6$ m s⁻¹
        (ii) 49 m s⁻¹
    (c) $v = \sqrt{200^2 + 49^2} = 205$ m s⁻¹
        $\tan \theta = 49/200$ $\theta = 13.8°$ to the horizontal (downwards)
    (d) $1500 = 0.5 \times 9.8 \times t^2$
        $t = 17.5$ s
    (e) 3500 m

12. (a) 0 m s⁻¹
    (b) 17.5 m s⁻¹
    (c) 9.8 m s⁻² downwards
    (d) −18.7 m s⁻¹
    (e) 25.6 m s⁻¹ downwards at 47° to the horizontal

13. total accelerating force = 1150 + 800 = 1950 N
    acceleration = $\dfrac{1950}{83}$ = 23.5 m s⁻²

14. $F = 550\,000 - 49\,000 = 60\,000$ N
    acceleration = 1.2 m s⁻² (allow for the weight of the rocket when finding the resultant force)

15. (a) net accelerating force = 8000 − 3000 = 5000 N
    (b) acceleration = $\dfrac{5000}{1800}$ = 2.78 m s⁻²
    (c) force = ma = 800 × 2.78 = 2222 N

16. impulse = Ft = area below line = 10.2 N s
    impulse = $m\Delta v$
    mass = 2.5 kg
    $\Delta v = \dfrac{10.2}{2.5} = 4.1$ m s⁻¹

17. Using momentum before collision = momentum after collision:
    $(0.200 \times 8) + (1.5 \times 0) = 1.7v$, so $1.6 = 1.7v$
    Therefore $v = 0.94$ m s⁻¹

18. (a) force = 500 cos 30 = 433 N
    (b) force = 500 sin 30 = 250 N
    (c) force = 433 N
    work done = 433 × 40 = 17 320 J

19. (a) power = Fv = 5000 × 30 = 150 kW
    (b) work done = Fs = 7000 × 35 = 245 000 J

20. energy change = 85 × 9.8 × 30 × 0.15 = 3750 J = 3.75 kJ

21. power = $2000 \times 10^3 = 250\,000 \times 9.8 \times \dfrac{v}{10}$
    Therefore, $v = 8.16$ m s⁻¹

22. $M = 400$ kg $m = 70$ kg
    (a) $F = ma + mg = 70(a + g)$
        $s = ut + \frac{1}{2}at^2$
        $a = 2 \times \dfrac{5}{16} = 0.625$ m s⁻²
        Therefore reading:
        $F = 70(9.8 + 0.625) = 70 \times 10.425 = 729.75$ N
    (b) The reading will be constant.
    (c) (i) $v^2 = u^2 + 2as$
        $= 2 \times 0.0.625 \times 5 = 6.25$ $v = 2.5$ m s⁻¹
        energy = gravitational potential energy + kinetic energy = $(5 \times 470 \times 9.8) + \frac{1}{2}(470 \times 2.5^2)$
        Therefore:
        total energy = 23 030 + 1468.75 = 24 498 J = 24.5 kJ
        (ii) This time there will be no kinetic energy increase.
        $s = 2.5 \times 4 = 10$ m
        Therefore:
        energy = 470 × 9.8 × 10 = 46 060 J = 46.1 kJ

23. density = $\dfrac{3 \times 10^{-6}}{3.75} = 8 \times 10^{-7}$ kg m⁻³

24. force = 50 = 9400 × e
    $e = 5.32$ mm
    new length = 2.50 + 0.0053 = 2.505 m

25. area = $\dfrac{force}{breaking\ stress} = \dfrac{30\,000}{500} \times 10^6 = 6 \times 10^{-5}$ m²
    (Note: 30 000 and not 20 000 to allow for the 50% safety margin.)
    Therefore each cable must have a cross-sectional area of 1.5×10^{-5} m².
    $1.5 \times 10^{-5} = \pi d^2/4$
    $d = 4.3 \times 10^{-3}$ m = 4.3 mm

26 additional energy $= \frac{1}{2}F\Delta L = 0.5 \times 70 \times 10^{-3} = 0.35\,J$

27 (a) $L = 0.1\,m$

$e = \frac{FL}{EA} = 0.05\,m$

energy $= \frac{1}{2}Fe = \frac{1}{2}Eae^2/L$

$= \frac{1}{2}(5 \times 10^8 \times 1 \times 10^{-6} \times 0.05^2)/0.5 = 6.25\,J$

(b) $\frac{1}{2}mv^2 = 6.25\,v = \sqrt{\frac{2 \times 6.25}{5 \times 10^{-3}}} = 50\,m\,s^{-1}$

(c) $v^2 = u^2 + 2as$

$50^2 = 2 \times 9.81 \times s$

$s = \frac{2500}{19.62} = 127\,m$

28 extension (e) $= \frac{FL}{EA}$

$= \frac{70 \times 9.81 \times 0.4 \times 5}{1.8 \times 10^{10} \times \pi \times (10^{-2})^2}$

$= 2.42 \times 10^{-4}\,m = 0.242\,mm$

29 stress $= \frac{force}{area} = \frac{70 \times 9.8}{\pi \times (1.5 \times 10^{-2})^2} = \frac{686}{7.068} \times 10^{-4}$

$= 9.70 \times 10^5\,Pa$

Chapter 5

1 (a) current $= \frac{20}{5} = 5\,A$

(b) $\frac{600}{20 \times 60} = 0.5\,A$

2 (a) charge $= It = 2 \times 10 = 20\,C$

(b) charge $= 5 \times 10^{-3} \times 8 \times 60 = 2.4\,C$

3 (a) $I = \frac{V}{R} = \frac{0.2}{120} = 1.7\,mA$

(b) $I = \frac{12}{4700} = 2.55\,mA$

(c) $I = \frac{6}{10^4} = 0.6\,mA$

(d) $I = \frac{25}{2.5} \times 10^6 = 10\,\mu A$

4 (a) $R = V/I = 12/0.25 = 48\,\Omega$

(b) $R = 230/10 = 23\,\Omega$

5 $\rho = RA/L = (3 \times \pi \times (0.28 \times 10^{-3})^2)/2.5$
$= 2.96 \times 10^{-7}\,\Omega\,m$

6 $R = \rho L/A = 0.65 \times 10^{-3}/1.5 \times 10^{-6} = 433\,\Omega$

7 When it is switched on due to the sudden expansion which gives the wire a 'thermal shock'.

8 (a) $200 + 400 = 600\,\Omega$

(b) $\frac{1}{R} = \frac{1}{400} + \frac{1}{200} = 0.0075\,R = 133\,\Omega$

(c) $R = \frac{750 \times 450}{750 + 450} = 281\,\Omega$

(d) $\frac{1}{R} = \frac{1}{500} + \frac{1}{500} + \frac{1}{500}$

$R = 167\,\Omega$

(e) $R = \frac{250 \times 200}{250 + 200} + 100 = 211\,\Omega$

(f) $\frac{1}{R} = \frac{1}{600} + \frac{1}{150} + \frac{1}{1200} = 109\,\Omega$

9 (a) $R = 50\,k\Omega$

Therefore:

$I = \frac{V}{R} = \frac{6}{50} \times 10^3 = 0.12\,mA$

(b) The resistance of the LDR will increase and so the current flowing from the cell will decrease.

(c) $I = \frac{6}{100} \times 10^3 = 0.06\,mA$ when the resistance of the LDR is infinite or when it is disconnected from the circuit.

10 (a) $V = 6\,V$

$I = \frac{6}{200} = 0.03\,A$

(b) $I = \frac{12}{200} = 0.06\,A$

11 (a) energy $= 1.5 \times 3000 = 4500\,J$

(b) energy $= 1.5 \times 200 \times 10^{-6} \times 2.5 \times 3600 = 2.7\,J$

12 energy $= 12 \times 200 \times 1.5 = 3600\,J$

13 power $= \frac{V^2}{R} = \frac{230^2}{50} = 1.058\,kW$

14 power loss $= I^2R = 4.5^2 \times 0.13 \times 0.75$
$= 1.97\,W$

15 (a) $I_3 = 0.096\,A$ flowing clockwise

(b) E is as shown
$E = (0.096 \times 150) - (0.04 \times 100) = 10.4\,V$

16 (a) $I_3 = 0.3\,A$ flowing anticlockwise. E must be reversed.

(b) $I_3 = 0.3\,A$ flowing clockwise. E is as shown.

17 $V_2 = \frac{R_2}{R_1 + R_2} \times V$

(a) $V_2 = \frac{200}{300} \times 12 = 8\,V$

(b) $V_2 = \frac{20}{45} \times 10 = 4.4\,V$

(c) $V_2 = \frac{200}{450} \times 6 = 2.7\,V$

18 (a) The output p.d. will fall.

(b) The output p.d. will rise.

19 The output p.d will fall due to the greater loss of energy within the cell due to the greater current flowing through its internal resistance.

20 The voltmeter has such a high resistance that when the voltmeter is connected to the cell on its own only $1.3/10$ MΩ $= 1.3 \times 10^{-7}$A $= 0.13\,\mu$A flows in the circuit.

The difference in the readings is because when the $20\,\Omega$ resistor is connected more current flows from the cell and so much more energy is lost within the internal resistance.

(a) current in the external resistor $= \dfrac{1.25}{20} = 0.0625$A

(b) Using $E = V + Ir$:

$1.3 = 1.25 + 0.0625r$

$\dfrac{1.3 - 1.25}{0.0625} = r$

internal resistance, $r = 0.8\,\Omega$

21 (a) (i) $I = \dfrac{E}{R + r}$

$I = \dfrac{2.5}{20.15} = 0.12$A

(ii) potential difference across the terminals $= IR = 0.12 \times 20 = 2.48$V

(iii) power loss within the cell $= I^2r = 0.0154 \times 0.15 = 0.0023$W

(b) (i) $I = \dfrac{E}{R + r}$

$I = \dfrac{2.5}{1.15} = 2.17$A

(ii) potential difference across the terminals $= IR = 2.18 \times 1 = 2.17$A

(iii) power loss within the cell $= I^2r = 4.73 \times 0.15 = 0.71$W

Chapter 6

1 (a) $T = \dfrac{2\pi r}{v} = 2 \times \pi \times \dfrac{3}{5} = 3.77$s

(b) $n = \dfrac{v}{2\pi r} = \dfrac{5}{2} \times \pi \times 3 = 0.27\,$s^{-1}

(c) angular velocity $(\omega) = \dfrac{v}{r} = \dfrac{5}{3} = 1.67$ radians s^{-1}

(d) 0 (linear acceleration $= \dfrac{v^2}{r} = 8.33\,$m s^{-2})

2 Constant speed and acceleration with constant magnitude (but varying direction). Note that the accelerating force is not constant because its direction is constantly changing.

3 (a) $F = \dfrac{mv^2}{r} = 4 \times \dfrac{16}{2} = 32$N

(b) $n = \dfrac{v}{2\pi r} = \dfrac{4}{2} \times \pi \times 2 = 0.32\,$s^{-1}

(c) At a tangent to the orbit

(d) $F = \dfrac{mv^2}{r} = 4 \times \dfrac{16}{1} = 64$N

4 (a) $a = \dfrac{v^2}{r} = \dfrac{\left(\frac{2\pi r}{T}\right)^2}{r} = \dfrac{\left(\frac{2\times\pi\times42\times10^6}{86400}\right)^2}{42\times10^6} = \dfrac{9.329\times10^6}{42\times10^6}$

$= 0.22\,$m s^{-2}

(b) Communication

5 (a) 0.03 m

(b) 3 s

(c) 0.33 s

(d) $x = 0.03\sin\dfrac{2\pi t}{3} = 0.03\sin\dfrac{2\pi 0.05}{3}$

$= 0.03\sin\dfrac{2\pi 0.5}{3} = 0.03\sin 60 = 0.026\,$m

Note the conversion of the radians into degrees in the sine function.

6 (a) a and c

(b) b

(c) a and c

(d) b

(e) a and c

7 $T = 2\pi\sqrt{\dfrac{10^{-2}}{g}} = 0.2$s

8 $T = 2\pi\sqrt{\dfrac{m}{k}}$

Therefore:

$m = \dfrac{kT^2}{4\pi^2} = \dfrac{2.5\times4}{4\pi^2} = 0.25$kg

9 period of the motion $(T) = 2\pi\sqrt{\dfrac{L}{g}}$

The value of g here is slightly less, so the period of the pendulum will increase and the clock loses time.

10 period of the motion $(T) = 2\pi\sqrt{\dfrac{L}{g}} = 2\pi\sqrt{\dfrac{0.25}{3.8}}$
$= 1.61$s

11 $\omega = \dfrac{2\pi}{T} = \dfrac{2\pi}{1.8} = 3.49$

amplitude $= 30$cm$/2 = 0.15$m

(a) maximum kinetic energy $= \frac{1}{2}m\omega^2A^2$
$= 0.5 \times 1.5 \times 3.49^2 \times 0.15^2 = 0.21$ J

(b) This will occur at the point of zero displacement — the centre of an oscillation.

12 Yes, resonance will result but the amplitude will build up more slowly than if the driving force was equal to the natural frequency.

13 Let the initial temperature of the rivet be θ.

heat energy lost by rivet $= 0.15 \times 385 \times (\theta - 35)$
$= 57.75\theta - 2021.25$ J

heat energy gained by water $= 0.25 \times 4200(35 - 16)$
$= 19\,950$ J

Therefore:

$57.75\theta - 2021.25 = 19\,950$

$\theta = 380.5$°C

14 The large specific heat capacity of water means that a large amount of heat is needed to heat the oceans by a significant amount. This means that the land near to the oceans also varies in temperature by only relatively small amounts.

AQA A-level Physics 219

15 thermal energy required $= 25 \times 10^{-3} \times 330\,000$
$+ 225 \times 10^{-3} \times 4200 \times 20 = 8250 + 1.89 \times 10^4$
$= 27\,150\,J$

electrical energy input $= 60 \times t$

Therefore:

time required $= \dfrac{2.715 \times 10^4}{60} = 453\,s$

16 The steam first condenses, liberating a large amount of heat energy due to latent heat. It then cools to the temperature of the body. The water simply cools.

17 (a) $p_1 V_1 = p_2 V_2$

$8 \times 3000 = 1 \times V_2$

So:

$V_2 = 3000 \times 8 = 24\,000\,cm^3$

However, this is not the volume of gas that *comes out* of the cylinder because there will still be $3000\,cm^3$ of gas left in it.

The volume of gas that comes out of the cylinder is therefore $24\,000 - 3000$
$= 21\,000\,cm^3$.

(b) $\dfrac{p_1}{T_1} = \dfrac{p_2}{T_2}$

$\dfrac{5 \times 10^4}{300} = \dfrac{p_2}{500}$

$p_2 = \dfrac{5 \times 10^4 \times 500}{300} = 8.3 \times 10^4\,Pa$

18 $\dfrac{V_1}{T_1} = \dfrac{V_2}{T_2}$

$\dfrac{4 \times 10^{-3}}{293} = \dfrac{V_2}{373}$

$V_2 = 5.1 \times 10^{-3}\,m^3$

19 $\dfrac{87.5 - 64}{100} = 0.235$

absolute zero $= \dfrac{-64}{0.235} = -272.3\,°C$

20 $pV = nRT$

$T = 273 + 27 = 300\,K$

$n = \dfrac{10^5 \times 5}{8.3 \times 300} = 200.8 = 201$ moles

21 $pV = nRT \qquad m = \rho V$

$V = \dfrac{0.04}{1.2}$ with $n = 1$

$p = \dfrac{RT}{V} = \dfrac{8.3 \times 353 \times 1.2}{0.04} = 8.79 \times 10^4\,Pa$

22 $pV = NkT$

$N = \dfrac{pV}{kT} = \dfrac{(1.01 \times 10^5 \times 5 \times 10^{-3}}{1.38 \times 10^{-23} \times 273.15} = \dfrac{505}{3.77 \times 10^{-21}}$
$= 1.34 \times 10^{23}$

23 Use $p = \tfrac{1}{3}\rho c^2$ to give $c^2 = 3P/\rho$

(a) $482\,m\,s^{-1}$

(b) $389\,m\,s^{-1}$

(c) $490\,m\,s^{-1}$

(d) $306\,m\,s^{-1}$

(e) $1826\,m\,s^{-1}$

24 Use: $PV = \tfrac{1}{3}mNv^2$ to give $N = \dfrac{3PV}{mv^2}$

$= \dfrac{3 \times 2 \times 10^5 \times 0.1}{3.5 \times 10^{-26} \times 3.025 \times 10^5}$.

So:

$N = \dfrac{6 \times 10^4}{1.059 \times 10^{-20}} = 5.67 \times 10^{24}$

25 average kinetic energy $= \tfrac{1}{2}mc^2 = \dfrac{3}{2} RT/N_A$

$= \dfrac{3 \times 8.3 \times 300}{2 \times 6.04 \times 10^{23}} = 6.2 \times 10^{-21}\,J$

Chapter 7

1 $F = \dfrac{GMm}{r^2}$

(a) $F = 1.3 \times 10^{-8}\,N$

(b) $F = 1.7 \times 10^{-10}\,N$

(c) $F = 1.5 \times 10^8\,N$

2 B $\quad \dfrac{-GMm}{r}$

3 If the meteoroid starts from infinity with zero velocity and is accelerated towards the Earth by the Earth's field alone then the greatest speed that it could have when it reached the Earth is the escape velocity of the Earth.

4 Diagram B from Figure 7.6. It must agree with the conditions mentioned on page 143.

5 distance around the orbit $= 2\pi R = 2\pi \times 6.6 \times 10^6$
$= 4.15 \times 10^7\,m$

Time in the orbit:

Can use $\dfrac{T^2}{R^3} = \dfrac{4\pi^2}{GM}$ and so $T^2 = \dfrac{R^3 \times 4\pi^2}{GM}$.

This gives $T = 5398\,s$.

Speed in orbit:

$v = \dfrac{distance}{time} = \dfrac{4.15 \times 10^7}{5398} = 7.7 \times 10^3\,m\,s^{-1}$

6 $F = eE = ma$

Therefore:

$a = \dfrac{eE}{m} = \dfrac{1.6 \times 10^{-19} \times 1000}{9 \times 10^{-31}} = 1.8 \times 10^{14}\,m\,s^{-2}$

The electron accelerates parallel to the field direction (positive to negative).

7 electrostatic potential $= \dfrac{1}{4\pi\varepsilon_0} \dfrac{Q}{d}$

$= \dfrac{8.99 \times 10^9 \times 0.5 \times 10^{-3}}{0.5} = 9.0 \times 10^6\,V$

8 Use $= \dfrac{Q}{V}$:

(a) $Q = CV = 5 \times 10^{-6} \times 6 = 30 \times 10^{-6} = 30\,\mu C$

(b) $V = \dfrac{Q}{C} = \dfrac{2.5 \times 10^{-8}}{400 \times 10^{-12}} = 63\,V$

9 $C = \dfrac{\varepsilon_0 A}{d} = \dfrac{8.85 \times 10^{-12} \times 25 \times 10^{-4}}{2 \times 10^{-4}}$
$= 1.11 \times 10^{-10}\,F = 1.11 \times 10^{-4}\,\mu F$

10   $C = \dfrac{\varepsilon_0 A}{d}$

area $(A) = \dfrac{Cd}{\varepsilon_0} = \dfrac{100 \times 10^{-6} \times 0.5 \times 10^{-3}}{8.85 \times 10^{-12}}$

$= 5.65 \times 10^3\,\text{m}^2$

11   capacitance $= \dfrac{8.85 \times 10^{-12} \times 3.5 \times 25 \times 10^{-4}}{10^{-4}}$

$= 7.8 \times 10^{-10} = 0.78\,\text{nF}$

12   energy $= \frac{1}{2}QV = 0.5 \times 5 \times 100 = 250\,\text{J}$

13   capacitor energy $= \frac{1}{2}CV^2 = \frac{1}{2}\dfrac{\varepsilon_0 A}{d}V^2$

$= 0.5 \times \left(\dfrac{8.84 \times 10^{-12} \times 1}{10^{-3}}\right)20^2 = 1.77 \times 10^{-6}\,\text{J}$

14   (a) Unchanged

(b) Unchanged

(c) Decreases

15   $V = V_0\left(1 - e^{-\frac{t}{RC}}\right)$

Therefore:

$e^{-\frac{t}{RC}} = \dfrac{V_0 - V}{V_0} = 0.25$

$e^{\frac{t}{RC}} = 4$

$\dfrac{t}{RC} = \ln 4 = 1.39$

Therefore:

$t = 1.39 \times 1000 \times 10^{-6} \times 2 \times 10^3 = 2.78\,\text{s}$

16   Use $V = V_0 e^{-\frac{t}{RC}}$

$V = 12 \times e^{-\frac{0.5}{1000 \times 0.001}} = 12e^{-0.5} = 12 \times 0.607 = 7.3\,\text{V}$

17   time constant $= RC = 10 \times 10^{-6} \times 1000 \times 10^3 = 10\,\text{s}$

18   $F = BIL = mg$

Therefore:

$I = \dfrac{mg}{BL} = \dfrac{0.003 \times 9.81}{1.2 \times 0.1} = 0.25\,\text{A}$

19   $\dfrac{q}{m} = \dfrac{v}{Br} = \dfrac{2.1 \times 10^7}{0.03 \times 0.004} = 1.75 \times 10^{11}\,\text{C kg}^{-1}$

20   The orbit radius increases.

21   flux linkage $= BAN\cos\theta$

$\cos\theta = \dfrac{0.15}{0.5 \times 50 \times 10^{-4} \times 100} = 0.6$

Therefore:

$\theta = 53°$

22   (a) flux linkage $= BAN = 0.05 \times 0.2^2 \times 1$

$= 2 \times 10^{-3}\,\text{Wb}$

(b) Axis of coil is now at $90° - 30° = 60°$ to the field.

change of flux $= 2 \times 10^{-3} - BAN\cos 60$

$= 1 \times 10^{-3}\,\text{Wb}$

23   magnitude of the induced emf = rate of change of flux linkage

time $(dt) = \dfrac{Nd\Phi}{\text{emf}} = \dfrac{200 \times 2.3}{50} = 9.2\,\text{s}$

24   emf generated $= BLv\sin\theta$

Therefore:

$v = \dfrac{\text{emf}}{BL} = \dfrac{10}{1.5 \times 2 \times \sin 25} = 7.9\,\text{m s}^{-1}$

25   (a) Position of minimum emf is when the plane of the coil is at right angles to the field.

(b) emf $= BAN\omega$

Therefore:

$\omega = \dfrac{\text{emf}}{BAN} = \dfrac{12}{0.2 \times \pi \times 0.08^2 \times 200} = \dfrac{12}{0.8}$

$= 15\,\text{radian s}^{-1} = 2.4\,\text{rev s}^{-1}$

26   $V = \dfrac{V_0}{\sqrt{2}} = 0.707v_0 = 0.707 \times 156 = 110\,\text{V}$

27   0

28   (a) $T = 14\,\text{ms}$

$f = \dfrac{1}{T} = 71.4\,\text{Hz}$

(b) amplitude $= 21.5\,\text{V}$

(c) voltage $= -12.5\,\text{V}$

29   (a) 0

(b) $V_s = 0.88\left(\dfrac{V_p N_s}{N_p}\right) = \dfrac{0.88 \times 25 \times 2000}{1000} = 44\,\text{V}$

30   Aluminium has a lower density $(2710\,\text{kg m}^{-3})$ than silver $(10\,500\,\text{kg m}^{-3})$ so the power lines would be lighter — an advantage. However, the resistivity of aluminium $(2.65 \times 10^{-8}\,\Omega\,\text{m})$ is higher than that of silver $(1.6 \times 10^{-8}\,\Omega\,\text{m})$ — a disadvantage. Finally, the cost of aluminium (£1.50 per kg) is also less than silver (£335 per kg) — another advantage. (Figures correct at the time of writing in early 2016.)

31   power loss per km $= 8 = 10^2 \times R$

Therefore:

$R = \dfrac{8}{100} = 0.08\,\Omega\,\text{km}^{-1}$

Chapter 8

1   B $^{241}_{95}\text{Am}$ alpha emitter — short range in air but highly ionising

2   (a) time elapsed $= 2020 - 2016 = 4$ years

$= 4 \times 3.15 \times 10^7 = 1.26 \times 10^8\,\text{s}$

$\lambda t = 7.85 \times 10^{-10} \times 1.26 \times 10^8 = 9.89 \times 10^{-2}$

$= 0.0989$

$A = A_0 e^{-\lambda t} = 185 \times 10^3 e^{-0.0989} = 185 \times 10^3 \times 0.906$

$= 168 \times 10^3 = 168\,\text{kBq}$

(b) time elapsed $= 2035 - 2016 = 19$ years

$= 19 \times 3.15 \times 10^7 = 5.985 \times 10^8\,\text{s}$

$\lambda t = 7.85 \times 10^{-10} \times 5.99 \times 10^8 = 9.89 \times 10^{-2}$

$= 0.470$

$A = A_0 e^{-\lambda t} = 185 \times 10^3 e^{-0.470} = 185 \times 10^3 \times 0.625$

$= 116 \times 10^3 = 116\,\text{kBq}$

3   $\lambda = 10\,\text{s}^{-1}$

$A = A_0 e^{-\lambda t}$

Therefore:

$$\ln\left(\frac{A_0}{A}\right) = \lambda t$$

$$t = \ln\frac{\left(\frac{A_0}{A}\right)}{\lambda} = \ln\frac{10^6}{10} = \frac{13.82}{10} = 1.4\,s$$

4 Use $A = A_0/2^n$:
   (a) time difference = 2016 – 1986 = 30 years
      half-lives passed = 1
      activity = $5\,kBq\,m^{-2}$
   (b) time difference = 2026 – 1986 = 40 years
      half-lives passed = 1.33
      activity = $\dfrac{10}{2^{1.33}} = 4\,kBq\,m^{-2}$

5 reading due to sample alone = 170 – 10 = 160 Bq
   (a) Two half-lives have passed.
      reading due to source = $\dfrac{160}{4} = 40$
      total reading = 40 + background = 50 Bq
   (b) 38.3 Bq (use $A = A_0/2^n$ plus the background count)
   (c) 30 Bq

6 Use $A = A_0/2^n$:
   Count rate = 45 – 2 = 43
   (a) A = 11.2 Bq
   (b) A = 2.4 Bq
   In (b) the count rate due to the source is much lower than the background count.

7 C

8

Isotope	Half life	Particle emitted
Polonium-218	3.1 minutes	Alpha
Lead-214	27 minutes	Beta
Bismuth-214	20 minutes	Beta
Polonium-214	$1.6 \times 10^{-4}\,s$	Alpha
Lead-210	19 years	Beta
Bismuth-210	5.0 days	Beta
Polonium-210	138 days	Alpha
Lead-206	Stable	–

9 $d_c = \left(\dfrac{1}{4\pi\varepsilon_0}\right)\dfrac{qQ}{E_\alpha}$

$$= \frac{8.99 \times 10^9 \times 2 \times 1.6 \times 10^{-19} \times 79 \times 1.6 \times 10^{-19}}{4.5 \times 10^6 \times 1.6 \times 10^{-19}}$$

$$= 5.1 \times 10^{-14}\,m = 51\,fm$$

10 number of protons = 26   mass of protons = 26 × 1.00728 = 26.1893 u

   number of neutrons = 31   mass of neutrons = 31 × 1.00867 = 31.2688 u

   mass of the nuclear constituents = 57.4581 u

(a) binding energy of the nucleus
   = 57.4581 – 56.9353 = 0.5228 u = 487 MeV
(b) binding energy per nucleon = 0.5228/57
   = 0.00917 u = 8.54 MeV

11 235 g of U-235 contain 6.02×10^{23} nuclei. Therefore 1 kg contains 2.56×10^{24} nuclei.

   mass 'converted' to energy due to fission of 1 kg = $0.1276 \times 2.56 \times 10^{24} = 3.27 \times 10^{23}\,u$

   energy available from fission = $3.27 \times 10^{23} \times 931$ = $3.04 \times 10^{26}\,MeV = 4.9 \times 10^{13}\,J$

12 In the D–D reaction there are only two neutrons to 'spread out' the repulsive strong nuclear force between the two protons, whereas in the D–T reaction there are three.

13 (a) 2D = 2 × 2.014102 u = 4.08204 u
      n = 1.008665 u ^{3}He = 3.016048 u
      mass defect = 0.003491 u = 3.25 MeV
      = $3.25 \times 1.6 \times 10^{-13}\,J = 5.2 \times 10^{-13}\,J$
   (b) 2D = 2 × 2.014102 u = 4.08204 u
      ^{1}H = 1.007273 u
      ^{3}H = 3.016049 u
      mass defect = 0.004882 u = 4.545 MeV
      = $4.545 \times 1.6 \times 10^{-13}\,J = 7.27 \times 10^{-13}\,J$
   (c) 2T = 2 × 3.016049 u = 6.032098
      2n = 2 × 1.008665 u
      ^{4}He = 4.002604 u
      mass defect = 0.012164 u = 11.3 MeV
      = $11.3 \times 1.6 \times 10^{-13}\,J = 1.8 \times 10^{-13}\,J$

14 mass difference = (239.060765 + 1.00867)
   – (133.9054 + 102.9266 + 3 × 1.00867)
   = 240.069435 – 239.85801 = 0.2114 u

   1 u is equivalent to 931 MeV or $931 \times 1.6 \times 10^{-13}\,J$.

   Therefore, the energy available from the reaction is $0.2114 \times 931 \times 1.6 \times 10^{-13} = 3.15 \times 10^{-11}\,J$.

   239 g of ^{239}Pu contains 6.02×10^{23} nuclei. Therefore, 1 kg contains 2.52×10^{24} nuclei.

   Therefore:

   energy available from the fission of 1 kg of plutonium-239 = $3.15 \times 10^{-11} \times 2.52 \times 10^{24}$ = $7.93 \times 10^{13}\,J$

Chapter 9

1 $M = \dfrac{f_0}{f_e}$

   eye lens focal length $(f_e) = \dfrac{f_0}{1000} = 0.15\,cm$

2 $l = r\theta$

   $\theta = \dfrac{3500}{400000} = 0.00875\,rad$

   $= 0.05°$

3 (a) $\theta = \dfrac{1.22\lambda}{D} = \dfrac{1.22 \times 400 \times 10^{-9}}{0.15} = 3.25 \times 10^{-6}\,\text{rad}$

1 radian = $\left(\dfrac{3600 \times 180}{\pi}\right)$ arc seconds

$3.25 \times 10^{-6}\,\text{rad} = 3.25 \times 10^{-6} \times \left(\dfrac{3600 \times 180}{\pi}\right)$

$= 0.671'$

(b) Atmospheric conditions; colours spread.

4 $2.51^n = 1000$

Therefore:

$n \log 2.51 = \log 1000$

So:

$n = 7.51$

5 (a) $-0.7 - M = 5 \log\left(\dfrac{20}{10}\right)$

$-0.7 - M = 1.51$

$M = -2.21$

(b) $m - 2.6 = 5 \log\left(\dfrac{5}{10}\right)$

$m = 2.6 + (-1.51) = +1.09$

(c) $-0.2 - 2 = 5 \log\left(\dfrac{d}{10}\right)$

$-2.2 = 5 \log\left(\dfrac{d}{10}\right)$

$d = 3.63\,\text{pc} = 11.8$ light years

6 power $= \sigma A T^4 = 10^{30} = 5.67 \times 10^{-8} \times 4\pi \times (10^{11})^2 \times T^4$

Therefore:

$T^4 = \dfrac{10^{30}}{5.67 \times 10^{-8} \times 4\pi \times 10^{22}} = 1.4035 \times 10^{14}$

So:

$T = 3442\,\text{K}$

7 For a black hole, $c = \left(\dfrac{2GM}{R_S}\right)^{1/2}$, where R_S is the Schwarzschild radius.

$M = \dfrac{c^2 \times R_S}{2G} = \dfrac{9 \times 10^{16} \times 2 \times 10^5}{2 \times 6.67 \times 10^{-11}} = \dfrac{1.8 \times 10^{22}}{1.33 \times 10^{-10}}$

$= 1.35 \times 10^{32}\,\text{kg}$

8 (a) Blue shifted — moving towards the Earth.

(b) $\lambda' = \lambda\left(\dfrac{1-v}{c}\right)$

$\lambda = \dfrac{600}{\frac{3\times10^8 - 5\times10^5}{3\times10^8}} = \dfrac{600}{0.998} = 601\,\text{nm}$

The line would have a longer wavelength if the galaxy was at rest relative to the Earth.

(c) 1 nm

9 radius of the observable universe (R)

$= \dfrac{v}{H} = \dfrac{3 \times 10^8}{2.32 \times 10^{-18}} = 1.29 \times 10^{26}\,\text{m}$

$= \dfrac{1.29 \times 10^{26}}{9.46 \times 10^{15}} = 1.37 \times 10^{10}$ light years

10 $H = 70\,\text{km s}^{-1}\,\text{Mpc}^{-1}$

Then:

critical density (ρ) $= \dfrac{3H^2}{8\pi G}$

$= \dfrac{(2.27 \times 10^{-18})^2 \times 3}{8 \times 3.14 \times 6.67 \times 10^{-11}}$

$= 9.22 \times 10^{-27}\,\text{kg m}^{-3}$

11 percentage change in observed intensities
= percentage change in area of the star visible

$= \dfrac{(R^2 - r^2) \times 100}{R^2} = \dfrac{R^2 - (0.01 \times R^2) \times 100}{R^2} = 1\%$

Units, useful formulae and mathematics

Quantities units and symbols

Quantity	Unit	Symbol
activity	becquerel	Bq
amount of substance	mole	mol
atomic mass unit	u	u
capacitance	farad	F
electric charge	coulomb	C
electric potential difference	volt	V
electric field strength	newton per coulomb	NC^{-1}
electromotive force	volt	V
electric current	ampere	A
electric resistance	ohm	Ω
gravitational field strength	newton per kilogram	Nkg^{-1}
resistivity	ohm-metre	Ωm
force	newton	N

Quantity	Unit	Symbol
frequency	hertz	Hz
length	metre	m
magnetic flux density	tesla	T
magnetic flux	weber	Wb
mass	kilogram	kg
moment of a force torque	newton-metre	Nm
momentum	newton-second	Ns
power	watt	W
pressure and stress	pascal	Pa
thermodynamic temperature	kelvin	K
time	second	s
work, energy	joule	J
Young modulus	pascal	Pa

Useful formulae

Absolute magnitude (M): $m - M = 5\log(d/10)$	Moment = force × perpendicular distance from pivot
Accelerated motion: $a = \dfrac{\text{change in speed}}{\text{time taken}}$ $v = u + at \quad s = ut + \frac{1}{2}at^2 \quad v^2 = u^2 + 2as$	Momentum = mass × velocity
Capacitance (parallel plate) $= \dfrac{A\varepsilon_0\varepsilon_r}{d}$	Photoelectric effect: $hf = \phi + \frac{1}{2}mv^2$
Capacitance energy $= \frac{1}{2}QV = \frac{1}{2}CV^2 = \frac{1}{2}\dfrac{Q^2}{C}$	Potential difference $= \dfrac{\Delta W}{Q} = \dfrac{\text{energy change}}{\text{charge}} = \dfrac{\text{power}}{I}$
Capacitance charge: $Q = Q_0(1 - e^{-1/RC})$	Power $= \dfrac{\text{work}}{\text{time}} = $ force × velocity
Capacitance discharge: $Q = Q_0e^{-1/RC}$	Power (electrical) $= VI = \dfrac{V^2}{R} = I^2R$
Centripetal force $= \dfrac{mv^2}{r} = m\omega^2r$	Nuclear radius $(R) = R_0A^{\frac{1}{3}}$
Current $(I) = \dfrac{\text{charge}}{\text{time}} = \dfrac{\Delta Q}{\Delta t}$	Pressure $= \dfrac{\text{force}}{\text{area}}$
Density $(\rho) = \dfrac{\text{mass}}{\text{volume}}$	Radioactive decay: (activity) $A = \Delta n/\Delta t = -\lambda n$
Diffraction grating maximum: $n\lambda = d\sin\theta$	Radioactive decay: $n = n_0e^{-\lambda t}$

Doppler effect $(z) = \dfrac{\Delta f}{f} = \dfrac{v}{c} \dfrac{\Delta \lambda}{\lambda} = \dfrac{v}{c}$ v for $\ll c$	Radioactivity: activity $= A_0\, e^{-\lambda t}$
Double slit interference maximum: fringe width $(w) = \dfrac{\lambda D}{s}$	Radioactivity: half life $(T_{\frac{1}{2}}) = \ln 2/\lambda$
Electrical energy $= \dfrac{V}{t}$	Rayleigh criterion: $\theta \approx \dfrac{\lambda}{D}$
Electric field strength $(E) = \dfrac{F}{Q}$ Electric field strength (E) (uniform field) $= \dfrac{V}{d}$ Electric field strength (E) (radial field) $= \dfrac{1}{4\pi\varepsilon_0}\left[\dfrac{Q}{r^2}\right]$	Refractive index $(n) = \dfrac{\sin i}{\sin r} = \dfrac{c_i}{c_r}$
Energy (E) = Planck constant $(h) \times$ frequency	Resistance $= \dfrac{\text{voltage}}{\text{current}} = \dfrac{V}{I}$
Force $= ma$	Series: $R = R_1 + R_2 + R_3$ etc. Parallel: $\dfrac{1}{R} = \dfrac{1}{R_1} + \dfrac{1}{R_2} + \dfrac{1}{R_3}$ etc.
Force between charges $= \dfrac{1}{4\pi\varepsilon_0}\left[\dfrac{Q_1 Q_2}{r^2}\right]$	Root mean square: $I_{rms} = \dfrac{I_0}{\sqrt{2}} = \dfrac{V_0}{\sqrt{2}}$
Force on a moving charge in a magnetic field $= BQv$	Simple harmonic motion acceleration $= -\omega^2 x$ $x = A\cos \omega t$ and $v = \pm\omega\sqrt{A^2 - x^2}$ Helical spring: $T = 2\pi\sqrt{m/k}$ Simple pendulum: $T = 2\pi\sqrt{L/g}$
Force on a current carrying wire in a magnetic field $= BIL$	Specific heat capacity (c): $Q = mc\Delta\theta$
Gravitational potential energy $= mg\Delta h$	Specific latent heat: $Q = mL$
Gravitational force $= \dfrac{Gm_1 m_2}{r^2}$	Stefan's law: $P = \sigma A T^4$
G and g (radial field): $g = \dfrac{GM}{r^2}$	Strain $(\varepsilon) = \dfrac{\text{extension}}{\text{original length}}$
Gravitational potential (radial field) $= \dfrac{-GM}{r}$	Stress $(\sigma) = \dfrac{\text{force}}{\text{area}}$
Hubble's law: $v = Hd$	Torque = force × perpendicular distance between forces
Ideal gas equation: $pV = nRT = NkT$	Transformer: $\dfrac{N_s}{N_p} = \dfrac{V_s}{V_p}$
Induced emf $= \dfrac{N\Delta\varphi}{\Delta t}$	Uniform velocity: speed $= \dfrac{\text{distance}}{\text{time}}$ $s = vt$
Induced emf in a coil $= BAN\omega \sin \omega t$	Wavelength of particle $(\lambda) = \dfrac{h}{mv}$
Kinetic energy $= \frac{1}{2}mv^2$	Wave speed = frequency × wavelength
Kinetic theory of gases: $pV = \frac{1}{3}mN(c_{rms})^2$	Weight $= mg$
Magnetic flux $(\varphi) = BA$	Wien's law: $\lambda_{max} T =$ constant
Magnification $= \dfrac{f_0}{f_e}$	Work = force × distance
Mass–energy: $\Delta E = \Delta mc^2$	Young modulus $= \dfrac{\text{stress}}{\text{strain}} = \dfrac{FL}{A\Delta L}$
Molecular kinetic energy: $\frac{1}{2}m(c_{rms})^2 = \dfrac{3}{2kT} = \dfrac{3RT}{2N_A}$	

Mathematics

Make sensible evaluations of numerical expressions using reasonable approximations such as $\pi = 3$.

Use scientific notation such as $1.6 \times 10^6 \times 2 \times 10^5 = 3.2 \times 10^{11}$.

Change the subject of an algebraic equation. If $A = B \times C$, then $C = A/B$ and so $B = A/C$.

Solve algebraic equations of the form: $ax^2 + bx + c = 0$ using the formula:

$$\text{roots of the equation} = \frac{-b \pm \sqrt{b^2 - 4ac}}{2a}$$

Graphs

Recognise the form of simple graphs: linear, x^2, $1/x$, $\log x$ and e^x.

Logarithms

Equivalent forms of the logarithms of ab, a/b, x^n and e^{kx}:

$$\log(ab) = \log a + \log b$$

$$\log(a/b) = \log a - \log b$$

$$\log(x^n) = n \log x$$

$$\ln(e^{kx}) = kx$$

log implies a number to the base ten and was written as $\log_{10}$. We now usually write $\log_{10}$ as lg.

ln implies a number to the base e and was written as $\log_e$.

Area of a circle $= \pi r^2$

Circumference of a circle $= 2\pi r$ (one dimension, r)

Surface area of a sphere $= 4\pi r^2$ (two dimensions, r to the power 2)

Volume of a sphere $= 4/3\pi r^3$ (three dimensions, r to the power 3)

Surface area of a cylinder $= 2\pi r^2 + 2\pi rL$

Volume of a cylinder $= \pi r^2 L$

Use and apply simple theorems such as Pythagoras'.

Trigonometry

$\sin A$ = opposite side/hypotenuse

$\cos A$ = adjacent side/hypotenuse

$\tan A$ = opposite side/adjacent side

Recall that, when θ tends to zero, $\sin\theta$ tends to θ^c, $\cos\theta$ tends to 1 and $\tan\theta$ tends to θ^c, if θ^c is the angle expressed in radians.

Understand the use of the area below a curve when this has a physical significance.

Understand the use of the slope of a tangent to a curve to express rate of change.

Useful numbers: e = 2.7183; 1 radian = 57.3°

Translate from degrees to radians and vice versa, where θ radians $= (2\pi/360)\theta°$.